Table of Contents

The construction of Cube root

(An addition in Euclidean Geometry)

Manoranjan ghoshal

Published by "Self-published via Amazon Kindle Direct Publishing (KDP)"

Printed and distributed by Amazon Kindle Direct Publishing

www.amazon.com[1]

Book Description:

The Construction of Cube Root by Manoranjan Ghoshal is a thought-provoking addition to the field of classical Euclidean geometry. This work confronts one of the most historically significant unsolved problems in ancient Greek mathematics—the Dalian problem, or the doubling of the cube. Long believed to be impossible with only a compass and straightedge, this problem involves constructing a line segment whose length is the cube root of two, thereby doubling the volume of a unit cube.

1. http://www.amazon.com

Ghoshal takes readers through the historical and philosophical origins of geometry, tracing its evolution from primitive cave drawings to the sophisticated rigor of classical Greek thought. The book revisits the ancient mathematical constraints that rendered certain geometric problems unsolvable under traditional rules. In particular, it focuses on the challenge of cube root construction, presenting an original geometrical theorem that he argues allows this elusive construction using classical tools.

The core of the book is the author's own contribution—a proposed theorem and geometrical method to construct the cube root of a given line segment length. Through step-by-step logical reasoning, visual aids, and application of trigonometric relationships, Ghoshal presents his case that the doubling cube problem is, contrary to historical consensus, solvable within the framework of Euclidean construction.

Designed with accessibility in mind, especially for students and educators, the book uses clear language and detailed illustrations. It offers a blend of mathematical theory, historical commentary, and pedagogical clarity, making it a unique resource for researchers, teachers, and geometry enthusiasts alike.

Table of contents

This book is dedicated to
My parents ....

Introduction

Construction problems in classical geometry kept bound the uses. So our need takes the alternative way to fulfill the problem.

Although we accept the way of alternative for unsolved the problem.

The doubling cube is one of the construction problem in classical geometry unsolved.

Although it has no alternative even complimentary method, our Mathematicians tried to solved, showed the way but no one can get the success.

Here in this book "the construction of cube root" is showing the exact method or construction procedure clearly.

For better understand of readers specifically school students using lucid and explicit English language.

The book will be helpful teacher's researcher's students for everyone.

I acknowledge with thanks and owe to different sources of reference, also thanks to Amazon (KDP) for their tireless efforts of publishing and distributing worldwide.

Yours
Manoranjan ghoshal

Chapter: 1

Origin of geometry

Geometry is structural science or painting, and our school geometry means mathematical structure. Although, school geometry are two types, classical geometry or Euclidean geometry and non-Euclidean geometry.

So the origin of geometry or structural science almost unknown, drawing of structure was started since in the stage of ape, when they lived in cave, their cave painting is bearing the evidence[1]. I think thereafter the painting is developed very slowly to mathematics drawing that is geometry.

Chapter: 2

Origin of classical geometry

Origin of geometry is very contradictable. I think the geometry started in the age of stone[2], when our ancestors were tried to shape as a weapon from an ordinary stone. So the actually time of origin has been abandoned in time line of earth. But now our readable geometry was started in oldest civilizations[3].

Chapter: 3

CONSTRUCTION PROBLEMS in classical geometry

Classical geometry or Euclidean geometry or school geometry have the boundary of construction, in Greek frameworks has shown the ruler and compass are only instruments to draw mathematical drawing[4].

Ruler can only join two points and extended it through as straight; compass can cut equal segments only.

I think construction problems is in classical geometry many, but here I am discussing three famous Greek construction problems only.

For this boundary of construction tools some famous geometrical constructions yet unsolved. Most three famous Greek construction problems were, 1. Angle trisection, 2. Doubling cube, 3. Squaring circle.

Angle trisection problem is solved by me (it is my claim) is showing in the book construction of angle trisection.

Squaring circle problem is unsolved now; Ramanian's squaring circle solution is approximately latest solution I think.

Now here I am showing the doubling cube or cube root construction procedure.

About problem:

Doubling the cube, also known as the **Dalian problem**, is an ancient[a] [2][62][3:9] geometric[4] problem. Given the edge[5] of a cube[6], the problem requires the construction of the edge of a second cube whose volume[7] is double that of the first. As with the related

---

2. https://en.m.wikipedia.org/wiki/Doubling_the_cube#cite_note-1
3. https://en.m.wikipedia.org/wiki/Doubling_the_cube#cite_note-Kern1934-2
4. https://en.m.wikipedia.org/wiki/Geometry
5. https://en.m.wikipedia.org/wiki/Edge_(geometry)
6. https://en.m.wikipedia.org/wiki/Cube
7. https://en.m.wikipedia.org/wiki/Volume

problems of squaring the circle[8] and trisecting the angle[9], doubling the cube is now known to be impossible to construct by using only a compass and straightedge[10], but even in ancient times solutions were known that employed other methods.

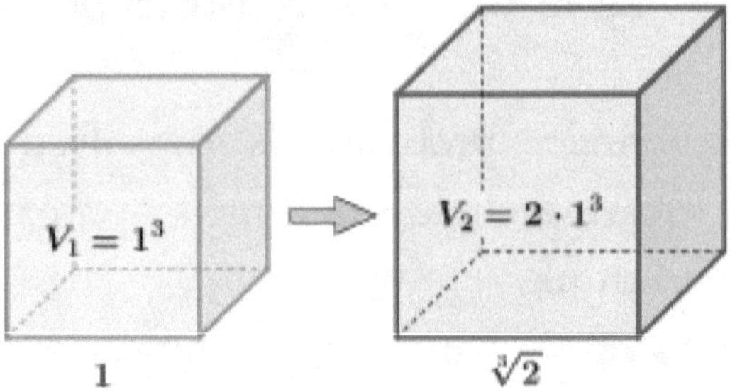

[11]A unit cube (side = 1) and a cube with twice the volume (side = cube root of two = 1.2599210498948732... OEIS[12]: A002580[13]).

According to Eutocius, Archytas was the first to solve the problem of doubling the cube (the so-called Dalian problem) with an ingenious geometric

8. https://en.m.wikipedia.org/wiki/Squaring_the_circle
9. https://en.m.wikipedia.org/wiki/Trisecting_the_angle
10. https://en.m.wikipedia.org/wiki/Compass_and_straightedge
11. https://en.m.wikipedia.org/wiki/File:01-W%C3%BCrfelverdoppelung-4.svg
12. https://en.m.wikipedia.org/wiki/On-Line_Encyclopedia_of_Integer_Sequences
13. https://oeis.org/A002580

construction.[63][14][64][15][65][16] The nonexistence of a compass-and-straightedge solution was finally proven by Pierre Wantzel[17] in 1837.

In algebraic terms, doubling a cube requires the construction of a line segment[18] of length x, where $x^3 = 2$; in other words,

the **cube root of two**. This is because a cube of side length 1 has a volume of $1^3 = 1$, and a cube of twice that volume (a volume of 2) has a side length of the cube root[19] of 2. The impossibility of doubling the cube is therefore equivalent[20] to the statement that cube root of two

is not a constructible number[21]. This is a consequence of the fact that the coordinates of a new point constructed by a compass and straightedge are roots of polynomials[22] over the field generated by the

14. https://en.m.wikipedia.org/wiki/Doubling_the_cube#cite_note-3
15. https://en.m.wikipedia.org/wiki/Doubling_the_cube#cite_note-4
16. https://en.m.wikipedia.org/wiki/Doubling_the_cube#cite_note-5
17. https://en.m.wikipedia.org/wiki/Pierre_Wantzel
18. https://en.m.wikipedia.org/wiki/Line_segment
19. https://en.m.wikipedia.org/wiki/Cube_root
20. https://en.m.wikipedia.org/wiki/Logical_equivalence
21. https://en.m.wikipedia.org/wiki/Constructible_number
22. https://en.m.wikipedia.org/wiki/Polynomial

coordinates of previous points, of no greater degree[23] than a quadratic[24]. This implies that the degree[25] of the field extension[26] generated by a constructible point must be a power of 2. The field extension generated by cube root of two however, is of degree 3.

ORIGIN OF PROBLEM:

The problem owes its name to a story concerning the citizens of Delos[27], who consulted the oracle at Delphi[28] in order to learn how to defeat a plague sent by Apollo[29].[66][30][62][31]:9 According to Plutarch[32],[67][33]

---

23. https://en.m.wikipedia.org/wiki/Degree_of_a_polynomial
24. https://en.m.wikipedia.org/wiki/Quadratic_polynomial
25. https://en.m.wikipedia.org/wiki/Degree_of_a_field_extension
26. https://en.m.wikipedia.org/wiki/Field_extension
27. https://en.m.wikipedia.org/wiki/Delos
28. https://en.m.wikipedia.org/wiki/Delphi
29. https://en.m.wikipedia.org/wiki/Apollo
30. https://en.m.wikipedia.org/wiki/Doubling_the_cube#cite_note-Zhmud-6
31. https://en.m.wikipedia.org/wiki/Doubling_the_cube#cite_note-Kern1934-2
32. https://en.m.wikipedia.org/wiki/Plutarch
33. https://en.m.wikipedia.org/wiki/Doubling_the_cube#cite_note-7

however, the citizens of Delos[34] consulted the oracle[35] at Delphi[36] to find a solution for their internal political problems at the time, which had intensified relationships among the citizens. The oracle responded that they must double the size of the altar to Apollo, which was a regular cube. The answer seemed strange to the Delian, and they consulted Plato[37], who was able to interpret the oracle as the mathematical problem of doubling the volume of a given cube, thus explaining the oracle as the advice of Apollo for the citizens of Delos[38] to occupy themselves with the study of geometry and mathematics in order to calm down their passions.[68][39]

According to Plutarch[40], Plato gave the problem to Eudoxus[41] and Archytas[42] and Menaechmus[43], who

---

34. https://en.m.wikipedia.org/wiki/Delos
35. https://en.m.wikipedia.org/wiki/Oracle
36. https://en.m.wikipedia.org/wiki/Delphi
37. https://en.m.wikipedia.org/wiki/Plato
38. https://en.m.wikipedia.org/wiki/Delos
39. https://en.m.wikipedia.org/wiki/Doubling_the_cube#cite_note-8
40. https://en.m.wikipedia.org/wiki/Plutarch
41. https://en.m.wikipedia.org/wiki/Eudoxus_of_Cnidus
42. https://en.m.wikipedia.org/wiki/Archytas
43. https://en.m.wikipedia.org/wiki/Menaechmus

solved the problem using mechanical means, earning a rebuke from Plato for not solving the problem using pure geometry[44].[69][45]This may be why the problem is referred to in the 350s BC by the author of the pseudo-Platonic *Sisyphus*[46] (388e) as still unsolved.[70][47]However another version of the story (attributed to Eratosthenes[48] by Eutocius of Ascalon[49]) says that all three found solutions but they were too abstract to be of practical value.[71][50]

A significant development in finding a solution to the problem was the discovery by Hippocrates of Chios[51] that it is equivalent to finding two geometric mean[52] proportional between a line segment and another with twice the length.[72][53] In modern notation, this means that given segments of lengths *a*

---

44. https://en.m.wikipedia.org/wiki/Pure_geometry
45. https://en.m.wikipedia.org/wiki/Doubling_the_cube#cite_note-9
46. *https://en.m.wikipedia.org/wiki/Sisyphus_(dialogue)*
47. https://en.m.wikipedia.org/wiki/Doubling_the_cube#cite_note-10
48. https://en.m.wikipedia.org/wiki/Eratosthenes
49. https://en.m.wikipedia.org/wiki/Eutocius_of_Ascalon
50. https://en.m.wikipedia.org/wiki/Doubling_the_cube#cite_note-11
51. https://en.m.wikipedia.org/wiki/Hippocrates_of_Chios
52. https://en.m.wikipedia.org/wiki/Geometric_mean

and 2*a*, the duplication of the cube is equivalent to finding segments of lengths *r* and *s* so that

a/r = r/s = s/2a, in turn, this means that, r = a * cube root of two

But Pierre Wantzel[54] proved in 1837 that the cube root[55] of 2 is not constructible[56]; that is, it cannot be constructed with straightedge and compass[57].[73][58]

---

53. https://en.m.wikipedia.org/wiki/Doubling_the_cube#cite_note-Heath-12
54. https://en.m.wikipedia.org/wiki/Pierre_Wantzel
55. https://en.m.wikipedia.org/wiki/Cube_root
56. https://en.m.wikipedia.org/wiki/Constructible_number
57. https://en.m.wikipedia.org/wiki/Straightedge_and_compass
58. https://en.m.wikipedia.org/wiki/Doubling_the_cube#cite_note-13

Chapter: 4

Problem impossibility

I think, the problem is unsolved because no one can produce a successful construction with ruler and compass.

Attempts:

Menaechmus' original solution involves the intersection of two conic[59] curves. Other more complicated methods of doubling the cube involve neusis[60], the cissoid of Diocles[61], the conchoid of Nicomedes[62], or the Philo line[63]. Pandrosion[64], a probably female mathematician of ancient Greece, found a numerically accurate approximate solution using planes in three dimensions, but was heavily criticized by Pappus of Alexandria[65] for not providing a proper mathematical proof[66].[74][67] Archytas[68] solved

59. https://en.m.wikipedia.org/wiki/Conic
60. https://en.m.wikipedia.org/wiki/Neusis
61. https://en.m.wikipedia.org/wiki/Cissoid_of_Diocles
62. https://en.m.wikipedia.org/wiki/Conchoid_(mathematics)
63. https://en.m.wikipedia.org/wiki/Philo_line
64. https://en.m.wikipedia.org/wiki/Pandrosion
65. https://en.m.wikipedia.org/wiki/Pappus_of_Alexandria
66. https://en.m.wikipedia.org/wiki/Mathematical_proof
67. https://en.m.wikipedia.org/wiki/Doubling_the_cube#cite_note-14

the problem in the 4th century BC using geometric construction in three dimensions, determining a certain point as the intersection of three surfaces of revolution.

Descartes theory of geometric solution of equations uses a parabola to introduce cubic equations, in this way it is possible to set up an equation whose solution is a cube root of two. Note that the parabola itself is not constructible except by three dimensional methods.

False claims of doubling the cube with compass and straightedge abound in mathematical crank[69] literature (pseudo mathematics).

Origami may also be used to construct the cube root of two by folding paper[70].

68. https://en.m.wikipedia.org/wiki/Archytas

69. https://en.m.wikipedia.org/wiki/Crank_(person)

70. https://en.m.wikipedia.org/wiki/Mathematics_of_paper_folding#Doubling_the_cube

Using a marked ruler

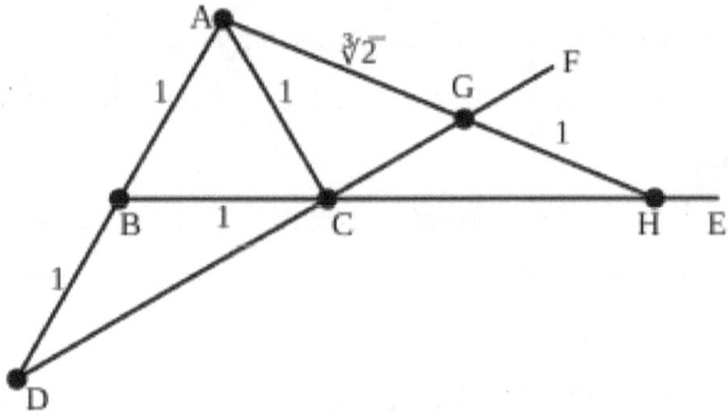

1

There is a simple neusis construction[2] using a marked ruler for a length which is the cube root of 2 times another length.[75][3]

Mark a ruler with the given length; this will eventually be GH.

Construct an equilateral triangle ABC with the given length as side.

Extend AB an equal amount again to D.

---

1. https://en.m.wikipedia.org/wiki/File:Doubling_the_cube.svg

2. https://en.m.wikipedia.org/wiki/Neusis_construction

3. https://en.m.wikipedia.org/wiki/Doubling_the_cube#cite_note-15

Extend the line BC forming the line CE.

Extend the line DC forming the line CF.

Place the marked ruler so it goes through A and one end, G, of the marked length falls on ray CF and the other end of the marked length, H, falls on ray CE. Thus GH is the given length.

Then AG is the given length times cube root of two.

Proof of impossibility:

We begin with the unit line segment defined by points[4] (0,0) and (1,0) in the plane[5]. We are required to construct a line segment defined by two points separated by a distance of cube root of two. It is easily shown that compass and straightedge constructions would allow such a line segment to be freely moved to touch the origin[6], parallel[7] with the unit line segment - so equivalently we may consider the task of constructing a line segment from (0,0) to (cube root of two, 0), which entails constructing the point (cube root of two, 0).

Respectively, the tools of a compass and straightedge allow us to create circles[8] centred[9] on one

4. https://en.m.wikipedia.org/wiki/Coordinate

5. https://en.m.wikipedia.org/wiki/Cartesian_plane

6. https://en.m.wikipedia.org/wiki/Origin_(mathematics)

7. https://en.m.wikipedia.org/wiki/Parallel_(geometry)

previously defined point and passing through another, and to create lines passing through two previously defined points. Any newly defined point either arises as the result of the intersection[10] of two such circles, as the intersection of a circle and a line, or as the intersection of two lines. An exercise of elementary analytic geometry[11] shows that in all three cases, both the *x*- and *y*-coordinates of the newly defined point satisfy a polynomial of degree no higher than a quadratic, with coefficients[12] that are additions, subtractions, multiplications, and divisions involving the coordinates of the previously defined points (and rational numbers). Restated in more abstract terminology, the new *x*- and *y*-coordinates have minimal polynomials[13] of degree at most 2 over the subfield[14] of R generated by the previous coordinates. Therefore, the degree[15] of the

---

8. https://en.m.wikipedia.org/wiki/Circles
9. https://en.m.wikipedia.org/wiki/Centre_(geometry)
10. https://en.m.wikipedia.org/wiki/Intersection
11. https://en.m.wikipedia.org/wiki/Analytic_geometry
12. https://en.m.wikipedia.org/wiki/Coefficients
13. https://en.m.wikipedia.org/wiki/Minimal_polynomial_(field_theory)
14. https://en.m.wikipedia.org/wiki/Field_(mathematics)
15. https://en.m.wikipedia.org/wiki/Degree_of_a_field_extension

field extension[16] corresponding to each new coordinate is 2 or 1.

So, given a coordinate of any constructed point, we may proceed inductively[17] backwards through the x- and y-coordinates of the points in the order that they were defined until we reach the original pair of points (0,0) and (1,0). As every field extension has degree 2 or 1, and as the field extension over Q of the coordinates of the original pair of points is clearly of degree 1, it follows from the tower rule[18] that the degree of the field extension over Q of any coordinate of a constructed point is a power of 2[19].

Now, $p(x) = x^3 - 2 = 0$ is easily seen to be irreducible[20] over Z[21] – any factorization would involve a linear factor[22] $(x - k)$ for some $k \in Z$, and so k must be a root[23] of $p(x)$; but also k must divide 2 (by the

---

16. https://en.m.wikipedia.org/wiki/Field_extension
17. https://en.m.wikipedia.org/wiki/Mathematical_induction
18. https://en.m.wikipedia.org/wiki/Degree_of_a_field_extension#The_multiplicitivity_formula_for_degrees
19. https://en.m.wikipedia.org/wiki/Power_of_2
20. https://en.m.wikipedia.org/wiki/Irreducible_polynomial
21. https://en.m.wikipedia.org/wiki/Integer
22. https://en.m.wikipedia.org/wiki/Linear_factor
23. https://en.m.wikipedia.org/wiki/Mathematics

rational root theorem[24]); that is, $k = 1, 2, -1$ or -2, and none of these are roots of $p(x)$. By Gauss's Lemma[25], $p(x)$ is also irreducible over Q, and is thus a minimal polynomial over Q for cube root of two. The field extension Q.(cube root of two) :Q is therefore of degree 3. But this is not a power of 2, so by the above: cube root of two is not the coordinate of a constructible point, so a line segment of cube root of two cannot be constructed with ruler and compass, and the cube cannot be doubled using only a ruler and a compass.

---

24. https://en.m.wikipedia.org/wiki/Rational_root_theorem

25. https://en.m.wikipedia.org/wiki/Gauss%27s_lemma_(polynomial)

Chapter: 5

My contribution

Limit of operation of ruler and compass never control the development of classical geometric construction. Here I am showing a new theorem and development of its application can construct the cube root length of an given length of line.

Doubling cube is an ancient famous classical construction problem in geometry [76,77,78,79,80,81,] is a part of cube root extraction. Unsuccessfully of geometers inspired to find out the impossibility. They produced the impossibility with algebraic geometry's cubic equation and number field concept. So, the problem has been accepted impossible.

Now I have find out a new theorem in classical geometry and it's development with application give us the result of cube root length of given length of line.

Using classical geometric construction tools, compass and ruler and a pencil also, with obey the Euclidean method and limit of operation of ruler and compass.

Theorem:

If two triangles have an equal angle, the ratio of opposite sides be never change with decrease or increase of angle equally.

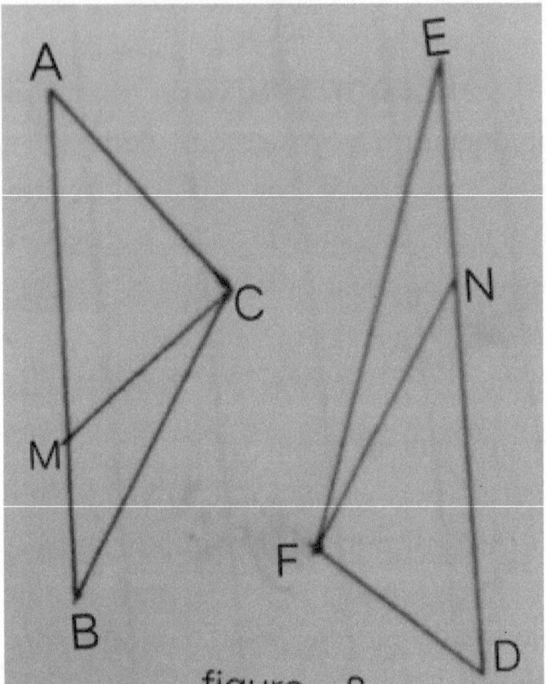

figure - 8

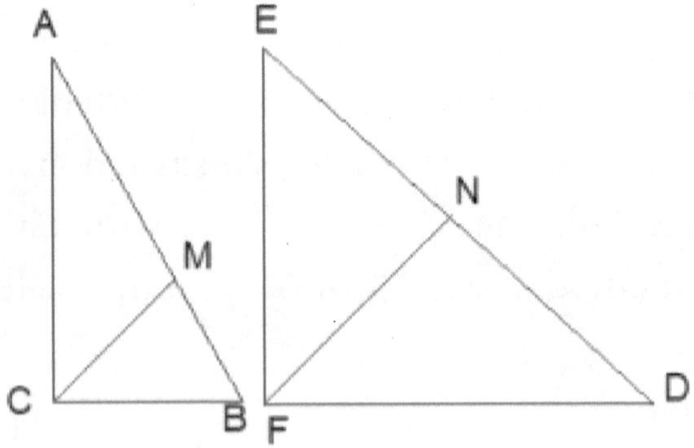

fig-8

in triangle ABC and triangle DEF has angle ACB = angle DFE, and angle BCM = angle EFN, I have need to prove, AB/DE = AM/DN.

According to "sine rule"

Sin ACB/AB,

And, sin DFE/DE

So, the ratio between them,

Sin ACB/AB: sin DFE/DE

Or, AB/DE: sin ACB/sin DFE

Or, AB/DE: 1,

After that,

Sin (ACB – BCM)/ (AB – BM),

And, sin (DFE – EFN)/(DE – EN)

Therefore, the ratio,

Sin ACM/AM: sin DFN/DN

Or, AM/DN: sin ACM/sin DFN

Or, AM/DN: 1

So, AB/DE = AM/DN (proved)

Now, Appling the theorem:

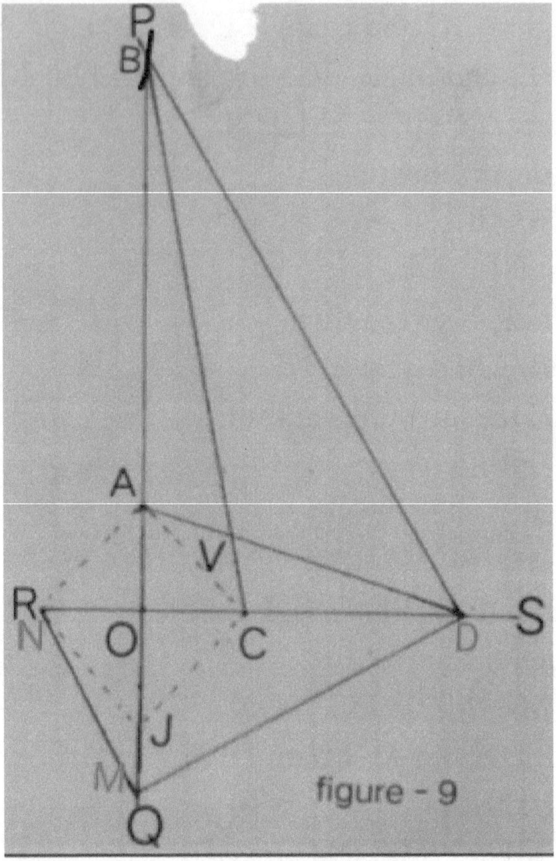

PQ and RS two straight line

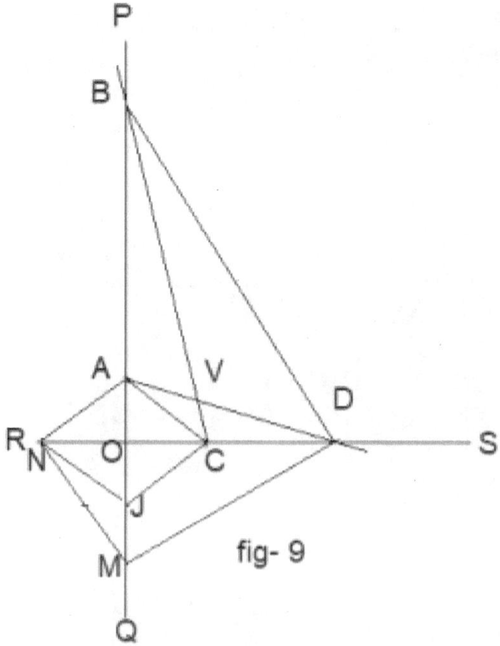

fig- 9

perpendicular situated at O, ACJR be

a square of unit length or one, triangle NMD and triangle MDB are right angled, angle NMD = angle BDM = 90°, (see fig – 9)

So, OB = $(OM)^3$,

And, OD = $(OM)^2$ are formed,

Here, sin AVB/AB: sin CVD/CD

Or, AB/CD: sin AVB/sin CVB

Or, AB/CD: 1,

Suppose, here OE be the length of given, (see fig – 10)

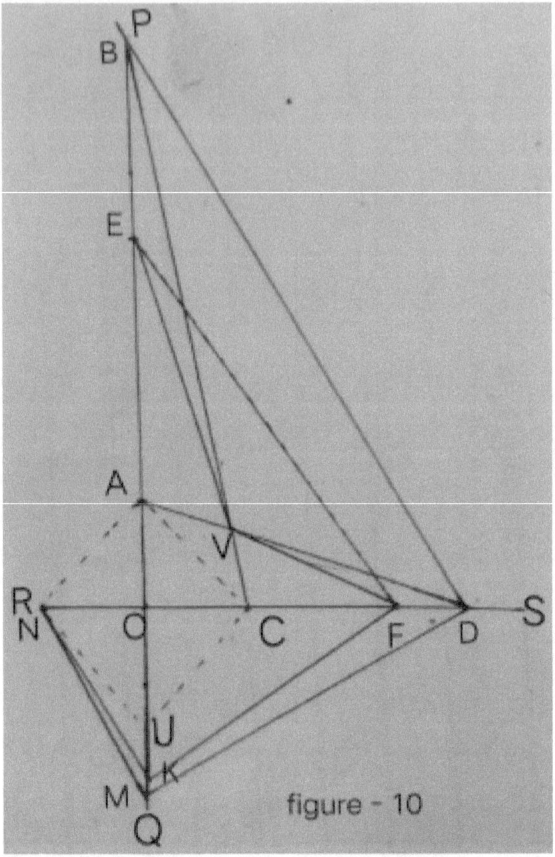

figure - 10

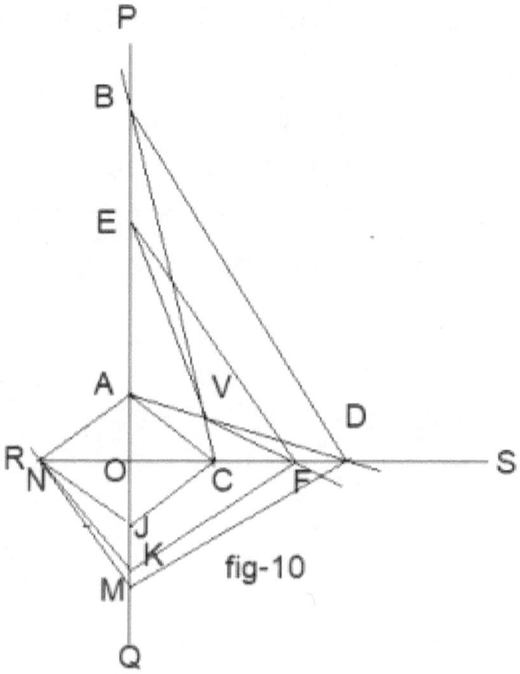

fig-10

Join E,V and set the angle DVF = angle BE, join E, F,

So, sin EVA/EA,

and, sin CVF/FC

Or, EA/FC : sin EVA/sin CVF

Or, EA/FC : 1

So, the ratio,

AB/CD = EA/FC

Therefore, the length (OK)3 = OE,

And, (OK)2 = OF,

Result:

Getting cube root length OK, OE be the given length of line.

Supplementary drawing:

Showing here the solution of most significant problem the doubling cube, and cube root extraction of given length of line is eight, graphically.

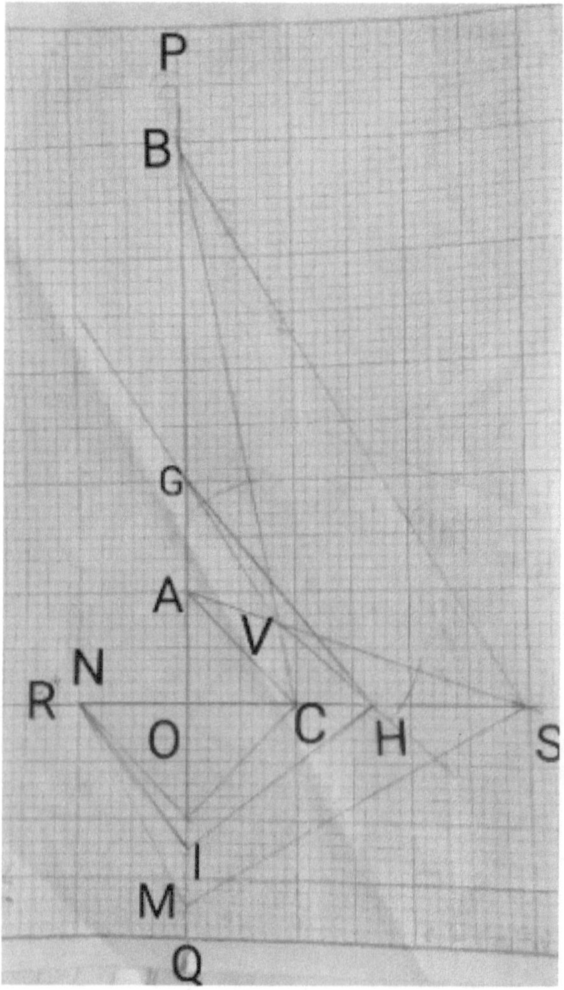
P
B
G
A
V
N
R
C
O
H
S
I
M
Q

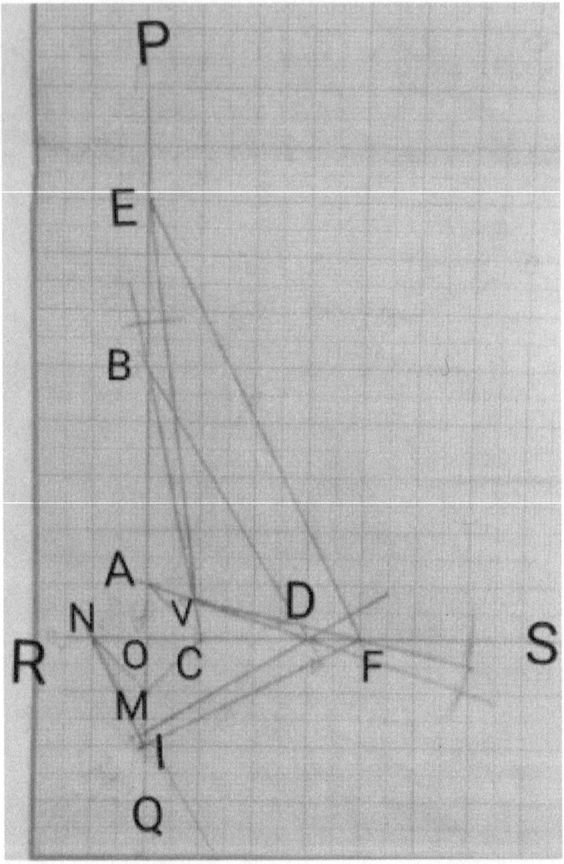

Extraction of doubling cube, here OI is the cube root of OG.

Extraction of cube root length of eight, here OI is the cube root length of OE.

Discussion:

Use of this method, cube root length construction possible of a given length of line in classical geometry with uses of ruler and compass.

Chapter: 6
Conclusion

Above of my discussion is clear, construction of cube root in classical geometry exactly possible. So, I can claim the doubling cube problem solved.

Selected references

1. ^[26] *Bruce Cole; Adelphi M. Gealt (15 December 1991). Art of the Western World: From Ancient Greece to Post Modernism[27]. Simon and Schuster. ISBN[28] 978-0-671-74728-2[29]. Retrieved 8 September 2011.*

2.^ ***a***[30] ***b***[31] *The Meeting of Eastern and Western Art*, Revised and Expanded edition (Hardcover) by Michael Sullivan[32].

3.^[33] *"Art View; Eastern Art Through Western Eyes"[34]. The New York Times[35]. 10 July 1994. Retrieved 30 November 2010.*

4.^ ***a***[36] ***b***[37] *Wichmann, Siegfried (1999). Japonisme: The Japanese Influence on Western Art since 1858. Thames & Hudson. ISBN[38] 978-0-500-28163-5[39].*

5.^ ***a***[40] ***b***[41] *Sullivan, Michael (1989). The Meeting of Eastern and Western Art[42]. University of California Press. ISBN[43] 978-0-520-05902-3[44].*

6.^[45] Discussion of the role of patrons in the Renaissance[46]. Retrieved 11 November 2008.

---

26. **https://en.m.wikipedia.org/wiki/History_of_painting#cite_ref-ColeGealt1991_1-0**

27. https://archive.org/details/artofwesternworl00bruc

28. https://en.m.wikipedia.org/wiki/ISBN_(identifier)

29. https://en.m.wikipedia.org/wiki/Special:BookSources/978-0-671-74728-2

30. https://en.m.wikipedia.org/wiki/History_of_painting#cite_ref-ReferenceA_2-0

31. https://en.m.wikipedia.org/wiki/History_of_painting#cite_ref-ReferenceA_2-1

32. https://en.m.wikipedia.org/wiki/Michael_Sullivan_(art_historian)

33. https://en.m.wikipedia.org/wiki/History_of_painting#cite_ref-3

34. https://query.nytimes.com/gst/fullpage.html?res=9C05EED91E3CF933A25754C0A962958260&sec=&spon=&pagewanted=2

35. https://en.m.wikipedia.org/wiki/The_New_York_Times

36. https://en.m.wikipedia.org/wiki/History_of_painting#cite_ref-Publisher_1999_4-0

37. https://en.m.wikipedia.org/wiki/History_of_painting#cite_ref-Publisher_1999_4-1

38. https://en.m.wikipedia.org/wiki/ISBN_(identifier)

39. https://en.m.wikipedia.org/wiki/Special:BookSources/978-0-500-28163-5

40. https://en.m.wikipedia.org/wiki/History_of_painting#cite_ref-Western_Art_1989_5-0

41. https://en.m.wikipedia.org/wiki/History_of_painting#cite_ref-Western_Art_1989_5-1

42. https://archive.org/details/bub_gb_PMFwC1gP0BkC

43. https://en.m.wikipedia.org/wiki/ISBN_(identifier)

44. https://en.m.wikipedia.org/wiki/Special:BookSources/978-0-520-05902-3

45. https://en.m.wikipedia.org/wiki/History_of_painting#cite_ref-6

46. https://web.archive.org/web/20010420173912/http:/www.geocities.com/rr17bb/patronage.html

7.^[47] History 1450–1789: Artistic Patronage[48]. Retrieved 11 November 2008.

8.^[49] Britannica.com[50]. Retrieved 11 November 2008.

9.^[51] Victorianweb.org[52], Aesthetes, Decadents, and the Idea of Art for Art's Sake George P. Landow, Professor of English and the History of Art, Brown University. Retrieved 11 November 2008.

10.^[53] *Cézanne to Picasso: Ambroise Vollard, Patron of the Avant-Garde*, Chicago Art Institute. Retrieved 11 November 2008[54]Archived[55] 12 October 2008 at the Wayback Machine[56].

13.^ ***a***57 ***b***58 *"Oldest tool use and meat-eating revealed | Natural History Museum"*59. 18 August 2010. Archived from *the original*60 on 18 August 2010.

14.^61 *"Stone Age"*62. HISTORY. *Retrieved 31 May 2020.*

---

47. **https://en.m.wikipedia.org/wiki/History_of_painting#cite_ref-7**

48. http://www.answers.com/topic/artistic-patronage

49. **https://en.m.wikipedia.org/wiki/History_of_painting#cite_ref-8**

50. https://www.britannica.com/eb/article-9125149/art-for-arts-sake

51. **https://en.m.wikipedia.org/wiki/History_of_painting#cite_ref-9**

52. http://www.victorianweb.org/decadence/artsake.html

53. **https://en.m.wikipedia.org/wiki/History_of_painting#cite_ref-10**

54. http://www.artic.edu/aic/exhibitions/picasso/themes.html

55. https://web.archive.org/web/20081012050647/http:/www.artic.edu/aic/exhibitions/picasso/themes.html

56. https://en.m.wikipedia.org/wiki/Wayback_Machine

57. ***https://en.m.wikipedia.org/wiki/Stone_Age#cite_ref-nhm.ac.uk_1-0***

58. ***https://en.m.wikipedia.org/wiki/Stone_Age#cite_ref-nhm.ac.uk_1-1***

59. *https://web.archive.org/web/20100818123718/http:/www.nhm.ac.uk/about-us/news/2010/august/oldest-tool-use-and-meat-eating-revealed75831.html*

60. *http://www.nhm.ac.uk/about-us/news/2010/august/oldest-tool-use-and-meat-eating-revealed75831.html*

61. **https://en.m.wikipedia.org/wiki/Stone_Age#cite_ref-2**

62. *https://www.history.com/topics/pre-history/stone-age*

15.^[63] Goody, Jack (2012). Metals, Culture, and Capitalism.

16.^[64] Ko, Kwang Hyun (2016). *"Origins of human intelligence: The chain of tool-making and brain evolution"*[65] *(PDF)*. Anthropological Notebooks. **22** (1): 5–22.

17.^[66] Barham & Mitchell 2008[67], p. 106

18.^[68] Barham & Mitchell 2008[69], p. 147

19.^[70] *Morelle, Rebecca*[71] (20 May 2015). *"Oldest tools pre-date first humans"*[72]. BBC News.

20.^ *a*[73] *b*[74] Harmand, Sonia; et al. (21 May 2015). "3.3-million-year-old stone tools from Lomekwi 3, West Turkana, and Kenya". Nature. **521**(7552): *310*–315. *Bibcode*[75]: *2015Natur.521..310H*[76]. *doi*[77]: *10.1038/nature14464*[78]. *PMID*[79] *25993961*[80]. *S2CID*[81] *120728582*[82].

---

63. **https://en.m.wikipedia.org/wiki/Stone_Age#cite_ref-3**
64. **https://en.m.wikipedia.org/wiki/Stone_Age#cite_ref-4**
65. *http://www.drustvo-antropologov.si/AN/PDF/2016_1/Anthropological_Notebooks_XXII_1_Ko.pdf*
66. **https://en.m.wikipedia.org/wiki/Stone_Age#cite_ref-5**
67. https://en.m.wikipedia.org/wiki/Stone_Age#CITEREFBarhamMitchell2008
68. **https://en.m.wikipedia.org/wiki/Stone_Age#cite_ref-6**
69. https://en.m.wikipedia.org/wiki/Stone_Age#CITEREFBarhamMitchell2008
70. **https://en.m.wikipedia.org/wiki/Stone_Age#cite_ref-7**
71. *https://en.m.wikipedia.org/wiki/Rebecca_Morelle*
72. *https://www.bbc.com/news/science-environment-32804177*
73. ***https://en.m.wikipedia.org/wiki/Stone_Age#cite_ref-Harmand_2015_8-0***
74. ***https://en.m.wikipedia.org/wiki/Stone_Age#cite_ref-Harmand_2015_8-1***
75. *https://en.m.wikipedia.org/wiki/Bibcode_(identifier)*
76. *https://ui.adsabs.harvard.edu/abs/2015Natur.521..310H*

21. ^[83] Rogers & Semaw 2009[84], pp. 162–163

22. ^[85] Rogers & Semaw 2009[86], p. 155

23. ^[87] As to whether *aethiopicus* is the genus *Australopithecus*[88] or the genus *Paranthropus*[89], broken out to include the more robust forms, anthropological opinion is divided and both usages occur in the professional sources.

24. ^[90] Rogers & Semaw 2009[91], p. 164

---

77. *https://en.m.wikipedia.org/wiki/Doi_(identifier)*

78. *https://doi.org/10.1038%2Fnature14464*

79. *https://en.m.wikipedia.org/wiki/PMID_(identifier)*

80. *https://pubmed.ncbi.nlm.nih.gov/25993961*

81. *https://en.m.wikipedia.org/wiki/S2CID_(identifier)*

82. *https://api.semanticscholar.org/CorpusID:1207285*

83. https://en.m.wikipedia.org/wiki/Stone_Age#cite_ref-9

84. https://en.m.wikipedia.org/wiki/Stone_Age#CITEREFRogersSemaw2009

85. https://en.m.wikipedia.org/wiki/Stone_Age#cite_ref-10

86. https://en.m.wikipedia.org/wiki/Stone_Age#CITEREFRogersSemaw2009

87. https://en.m.wikipedia.org/wiki/Stone_Age#cite_ref-11

88. *https://en.m.wikipedia.org/wiki/Australopithecus*

89. *https://en.m.wikipedia.org/wiki/Paranthropus*

90. https://en.m.wikipedia.org/wiki/Stone_Age#cite_ref-12

91. https://en.m.wikipedia.org/wiki/Stone_Age#CITEREFRogersSemaw2009

25.^92 *Zimmer, Carl*93 (11 July 2018). *"Archaeologists in China Discover the Oldest Stone Tools Outside Africa"*94. *The New York Times*95. *Retrieved 13 July 2018.*

26.^[96] *Friberg, Jöran (1981). "Methods and traditions of Babylonian mathematics"*[97]*. Historia Mathematica.* ***8*** *(3): 277–318. doi*[98]*:10.1016/0315-0860(81)90069-0*[99]*.*

27.^[100] *Neugebauer, Otto*[101] *(1969) [1957]. "Chap. IV Egyptian Mathematics and Astronomy"*[102]*. The Exact Sciences in Antiquity (2 ed.). Dover Publications*[103]*. pp. 71–96. ISBN*[104] *978-0-486-22332-2*[105]*. Archived*[106] *from the original on 14 August 2020. Retrieved 27 February 2021..*

28.^[107] (Boyer 1991[108], "Egypt" p. 19)

29.^[109] *Ossendrijver, Mathieu (29 January 2016). "Ancient Babylonian astronomers calculated Jupiter's position from the area under a time-velocity graph". Science.* ***351*** *(6272): 482–484. Bibcode*[110]*:2016Sci...351..482O*[111]*.*

92. https://en.m.wikipedia.org/wiki/Stone_Age#cite_ref-NYT-20180711cz_13-0

93. *https://en.m.wikipedia.org/wiki/Carl_Zimmer*

94. *https://www.nytimes.com/2018/07/11/science/hominins-tools-china.html*

95. *https://en.m.wikipedia.org/wiki/The_New_York_Times*

96. https://en.m.wikipedia.org/wiki/Geometry#cite_ref-6

97. *https://doi.org/10.1016%2F0315-0860%2881%2990069-0*

98. *https://en.m.wikipedia.org/wiki/Doi_(identifier)*

99. *https://doi.org/10.1016%2F0315-0860%2881%2990069-0*

100. https://en.m.wikipedia.org/wiki/Geometry#cite_ref-7

101. *https://en.m.wikipedia.org/wiki/Otto_E._Neugebauer*

102. *https://books.google.com/books?id=JVhTtVA2zr8C&pg=PA71*

103. *https://en.m.wikipedia.org/wiki/Dover_Publications*

104. *https://en.m.wikipedia.org/wiki/ISBN_(identifier)*

105. *https://en.m.wikipedia.org/wiki/Special:BookSources/978-0-486-22332-2*

106. *https://web.archive.org/web/20200814151056/https:/books.google.com/books?id=JVhTtVA2zr8C*

107. https://en.m.wikipedia.org/wiki/Geometry#cite_ref-Boyer_1991_loc_43ec3e5dee6e706af7766fffea512721_Egypt_p._19_8-0

108. https://en.m.wikipedia.org/wiki/Geometry#CITEREFBoyer1991

109. https://en.m.wikipedia.org/wiki/Geometry#cite_ref-9

doi[112]*:10.1126/science.aad8085*[113]*. PMID*[114] *26823423*[115]*. S2CID*[116] *206644971*[117]*.*

30.^[118] *Depuydt, Leo (1 January 1998). "Gnomons at Meroë and Early Trigonometry". The Journal of Egyptian Archaeology.* ***84****: 171–180. doi*[119]*:10.2307/3822211*[120]*. JSTOR*[121] *3822211*[122]*.*

31.^[123] *Slayman, Andrew (27 May 1998). "Neolithic Skywatchers"*[124]*. Archaeology Magazine Archive. Archived*[125] *from the original on 5 June 2011. Retrieved 17 April 2011.*

32.^[126] (Boyer 1991[127], "Ionia and the Pythagoreans" p. 43)

33.^[128] Eves, Howard, *An Introduction to the History of Mathematics*[129], Saunders, 1990, ISBN[130] 0-03-029558-0[131].

---

110. *https://en.m.wikipedia.org/wiki/Bibcode_(identifier)*

111. *https://ui.adsabs.harvard.edu/abs/2016Sci...351..482O*

112. *https://en.m.wikipedia.org/wiki/Doi_(identifier)*

113. *https://doi.org/10.1126%2Fscience.aad8085*

114. *https://en.m.wikipedia.org/wiki/PMID_(identifier)*

115. *https://pubmed.ncbi.nlm.nih.gov/26823423*

116. *https://en.m.wikipedia.org/wiki/S2CID_(identifier)*

117. *https://api.semanticscholar.org/CorpusID:206644971*

118. https://en.m.wikipedia.org/wiki/Geometry#cite_ref-10

119. *https://en.m.wikipedia.org/wiki/Doi_(identifier)*

120. *https://doi.org/10.2307%2F3822211*

121. *https://en.m.wikipedia.org/wiki/JSTOR_(identifier)*

122. *https://www.jstor.org/stable/3822211*

123. https://en.m.wikipedia.org/wiki/Geometry#cite_ref-11

124. *http://www.archaeology.org/online/news/nubia.html*

125. *https://web.archive.org/web/20110605234044/http:/www.archaeology.org/online/news/nubia.html*

126. https://en.m.wikipedia.org/wiki/Geometry#cite_ref-Boyer_1991_loc_43ec3e5dee6e706af7766fffea512721_Ionia_and_the_Pythagoreans_p._43_12-0

127. https://en.m.wikipedia.org/wiki/Geometry#CITEREFBoyer1991

128. https://en.m.wikipedia.org/wiki/Geometry#cite_ref-13

129. *https://archive.org/details/introductiontohi0000eves*

130. https://en.m.wikipedia.org/wiki/ISBN_(identifier)

131. https://en.m.wikipedia.org/wiki/Special:BookSources/0-03-029558-0

34.^[132] *Kurt Von Fritz (1945). "The Discovery of Incommensurability by Hippasus of Metapontum". Classics in the History of Greek Mathematics. Annals of Mathematics; Boston Studies in the Philosophy of Science. Vol. 240. Annals of Mathematics, Trustees of Princeton University on Behalf of the Annals of Mathematics, Mathematics Department, Princeton University. pp. 211–231. doi*[133]*:10.1007/978-1-4020-2640-9_11*[134]*. ISBN*[135] *978-90-481-5850-8*[136]*. JSTOR*[137] *1969021*[138]*.*

35.^[139] *James R. Choike (1980). "The Pentagram and the Discovery of an Irrational Number"*[140]*. The Two-Year College Mathematics Journal.* ***11*** *(5): 312–316. doi*[141]*:10.2307/3026893*[142]*. JSTOR*[143] *3026893*[144]*. Archived*[145] *from the original on 9 September 2022. Retrieved 9 September 2022.*

36.^[146] (Boyer 1991[147], "The Age of Plato and Aristotle" p. 92)

37.^[148] (Boyer 1991[149], "Euclid of Alexandria" p. 119)

38.^[150] (Boyer 1991[151], "Euclid of Alexandria" p. 104)

---

132. https://en.m.wikipedia.org/wiki/Geometry#cite_ref-14

133. https://en.m.wikipedia.org/wiki/Doi_(identifier)

134. https://doi.org/10.1007%2F978-1-4020-2640-9_11

135. https://en.m.wikipedia.org/wiki/ISBN_(identifier)

136. https://en.m.wikipedia.org/wiki/Special:BookSources/978-90-481-5850-8

137. https://en.m.wikipedia.org/wiki/JSTOR_(identifier)

138. https://www.jstor.org/stable/1969021

139. https://en.m.wikipedia.org/wiki/Geometry#cite_ref-15

140. https://www.tandfonline.com/doi/abs/10.1080/00494925.1980.11972468

141. https://en.m.wikipedia.org/wiki/Doi_(identifier)

142. https://doi.org/10.2307%2F3026893

143. https://en.m.wikipedia.org/wiki/JSTOR_(identifier)

144. https://www.jstor.org/stable/3026893

145. https://web.archive.org/web/20220909203418/https:/www.tandfonline.com/doi/abs/10.1080/00494925.1980.11972468

146. https://en.m.wikipedia.org/wiki/Geometry#cite_ref-16

147. https://en.m.wikipedia.org/wiki/Geometry#CITEREFBoyer1991

148. https://en.m.wikipedia.org/wiki/Geometry#cite_ref-17

149. https://en.m.wikipedia.org/wiki/Geometry#CITEREFBoyer1991

150. https://en.m.wikipedia.org/wiki/Geometry#cite_ref-Boyer_1991_loc_43ec3e5dee6e706af7766fffea512721_Euclid_of_Alexandria_p._104_18-0

39.^[152] Howard Eves[153], *An Introduction to the History of Mathematics*[154], Saunders, 1990, ISBN[155] 0-03-029558-0[156] p. 141: "No work, except The Bible[157], has been more widely used...."

40.^[158] *O'Connor, J.J.; Robertson, E.F. (February 1996). "A history of calculus"*[159]*. University of St Andrews*[160]*. Archived from the original*[161] *on 15 July 2007. Retrieved 7 August 2007.*

41.^[162] *Staal, Frits*[163] *(1999). "Greek and Vedic Geometry". Journal of Indian Philosophy.* ***27****(1–2): 105–127. doi*[164]*:10.1023/A:1004364417713*[165]*. S2CID*[166] *170894641*[167]*.*

42.^[168] (Cooke 2005[169], p. 198): "The arithmetic content of the *Śulva Sūtras* consists of rules for finding Pythagorean triples such as (3, 4, 5), (5, 12, 13), (8, 15, 17), and (12, 35, 37). It is not certain what practical use these arithmetic rules had. The best conjecture is that they were part of religious ritual. A Hindu home was required to have three fires burning at three different altars. The three altars were to be of different shapes, but all three were to have the same area. These conditions led to certain "Diophantine"

---

151. https://en.m.wikipedia.org/wiki/Geometry#CITEREFBoyer1991

152. https://en.m.wikipedia.org/wiki/Geometry#cite_ref-19

153. https://en.m.wikipedia.org/wiki/Howard_Eves

154. https://archive.org/details/introductiontohi0000eves

155. https://en.m.wikipedia.org/wiki/ISBN_(identifier)

156. https://en.m.wikipedia.org/wiki/Special:BookSources/0-03-029558-0

157. https://en.m.wikipedia.org/wiki/The_Bible

158. https://en.m.wikipedia.org/wiki/Geometry#cite_ref-20

159. https://web.archive.org/web/20070715191704/http:/www-groups.dcs.st-and.ac.uk/~history/HistTopics/The_rise_of_calculus.html

160. https://en.m.wikipedia.org/wiki/University_of_St_Andrews

161. http://www-groups.dcs.st-and.ac.uk/~history/HistTopics/The_rise_of_calculus.html

162. https://en.m.wikipedia.org/wiki/Geometry#cite_ref-Staal_1999_21-0

163. https://en.m.wikipedia.org/wiki/Frits_Staal

164. https://en.m.wikipedia.org/wiki/Doi_(identifier)

165. https://doi.org/10.1023%2FA%3A1004364417713

166. https://en.m.wikipedia.org/wiki/S2CID_(identifier)

167. https://api.semanticscholar.org/CorpusID:170894641

168. https://en.m.wikipedia.org/wiki/Geometry#cite_ref-cooke198_23-0

169. https://en.m.wikipedia.org/wiki/Geometry#CITEREFCooke2005

problems, a particular case of which is the generation of Pythagorean triples, so as to make one square integer equal to the sum of two others."

43.^[170] (Hayashi 2005[171], p. 371)

44.^ ***a***[172] ***b***[173] (Hayashi 2003[174], pp. 121–122)

45.^[175] *Rāshid, Rushdī (1994). The development of Arabic mathematics : between arithmetic and algebra[176]. Boston Studies in the Philosophy of Science. Vol. 156. p. 35. doi[177]:10.1007/978-94-017-3274-1[178]. ISBN[179] 978-0-7923-2565-9[180]. OCLC[181] 29181926[182].*

46.^[183] (Boyer 1991[184], "The Arabic Hegemony" pp. 241–242) "Omar Khayyam (c. 1050–1123), the "tent-maker," wrote an *Algebra* that went beyond that of al-Khwarizmi to include equations of third degree. Like his Arab predecessors, Omar Khayyam provided for quadratic equations both arithmetic and geometric solutions; for general cubic equations, he believed (mistakenly, as the 16th century later showed), arithmetic solutions were impossible; hence he gave only geometric solutions. The scheme of using intersecting conics to solve cubics had been used earlier by Menaechmus, Archimedes, and Alhazan, but Omar Khayyam took the praiseworthy step of generalizing the method to cover all third-degree equations (having positive roots). .. For equations of higher degree than three, Omar Khayyam evidently

---

170. https://en.m.wikipedia.org/wiki/Geometry#cite_ref-hayashi2005-371_24-0

171. https://en.m.wikipedia.org/wiki/Geometry#CITEREFHayashi2005

172. https://en.m.wikipedia.org/wiki/Geometry#cite_ref-hayashi2003-p121-122_25-0

173. https://en.m.wikipedia.org/wiki/Geometry#cite_ref-hayashi2003-p121-122_25-1

174. https://en.m.wikipedia.org/wiki/Geometry#CITEREFHayashi2003

175. https://en.m.wikipedia.org/wiki/Geometry#cite_ref-26

176. https://archive.org/details/RoshdiRashedauth.TheDevelopmentOfArabicMathematicsBetweenArithmeticAndAlgebraSpringerNetherlands1994/page/n43/mode/2up

177. https://en.m.wikipedia.org/wiki/Doi_(identifier)

178. https://doi.org/10.1007%2F978-94-017-3274-1

179. https://en.m.wikipedia.org/wiki/ISBN_(identifier)

180. https://en.m.wikipedia.org/wiki/Special:BookSources/978-0-7923-2565-9

181. https://en.m.wikipedia.org/wiki/OCLC_(identifier)

182. https://search.worldcat.org/oclc/29181926

183. https://en.m.wikipedia.org/wiki/Geometry#cite_ref-27

184. https://en.m.wikipedia.org/wiki/Geometry#CITEREFBoyer1991

did not envision similar geometric methods, for space does not contain more than three dimensions, ... One of the most fruitful contributions of Arabic eclecticism was the tendency to close the gap between numerical and geometric algebra. The decisive step in this direction came much later with Descartes, but Omar Khayyam was moving in this direction when he wrote, "Whoever thinks algebra is a trick in obtaining unknowns has thought it in vain. No attention should be paid to the fact that algebra and geometry are different in appearance. Algebras are geometric facts which are proved."".

47.^[185] *O'Connor, John J.; Robertson, Edmund F.*[186] *"Al-Mahani"*[187]*. MacTutor History of Mathematics Archive*[188]*. University of St Andrews*[189]*.*

48.^[190] *O'Connor, John J.; Robertson, Edmund F.*[191] *"Al-Sabi Thabit ibn Qurra al-Harrani"*[192]*. MacTutor History of Mathematics Archive*[193]*. University of St Andrews*[194]*.*

49.^[195] *O'Connor, John J.; Robertson, Edmund F.*[196]*"Omar Khayyam"*[197]*. MacTutor History of Mathematics Archive*[198]*. University of St Andrews*[199]*.*

50.^[200] Boris A. Rosenfeld and Adolf P. Youschkevitch (1996), "Geometry", in Roshdi Rashed, ed., *Encyclopedia of the History of Arabic Science*[201], Vol. 2, pp. 447–494 [470], Routledge[202], London and New York:

---

185. **https://en.m.wikipedia.org/wiki/Geometry#cite_ref-28**

186. *https://en.m.wikipedia.org/wiki/Edmund_F._Robertson*

187. *https://mathshistory.st-andrews.ac.uk/Biographies/Al-Mahani.html*

188. *https://en.m.wikipedia.org/wiki/MacTutor_History_of_Mathematics_Archive*

189. *https://en.m.wikipedia.org/wiki/University_of_St_Andrews*

190. **https://en.m.wikipedia.org/wiki/Geometry#cite_ref-ReferenceA_29-0**

191. *https://en.m.wikipedia.org/wiki/Edmund_F._Robertson*

192. *https://mathshistory.st-andrews.ac.uk/Biographies/Thabit.html*

193. *https://en.m.wikipedia.org/wiki/MacTutor_History_of_Mathematics_Archive*

194. *https://en.m.wikipedia.org/wiki/University_of_St_Andrews*

195. **https://en.m.wikipedia.org/wiki/Geometry#cite_ref-30**

196. *https://en.m.wikipedia.org/wiki/Edmund_F._Robertson*

197. *https://mathshistory.st-andrews.ac.uk/Biographies/Khayyam.html*

198. *https://en.m.wikipedia.org/wiki/MacTutor_History_of_Mathematics_Archive*

199. *https://en.m.wikipedia.org/wiki/University_of_St_Andrews*

200. **https://en.m.wikipedia.org/wiki/Geometry#cite_ref-31**

201. *https://en.m.wikipedia.org/wiki/Encyclopedia_of_the_History_of_Arabic_Science*

202. https://en.m.wikipedia.org/wiki/Routledge

"Three scientists, Ibn al-Haytham, Khayyam, and al-Tusi, had made the most considerable contribution to this branch of geometry whose importance came to be completely recognized only in the 19th century. In essence, their propositions concerning the properties of quadrangles which they considered, assuming that some of the angles of these figures were acute of obtuse embodied the first few theorems of the hyperbolic and the elliptic geometries. Their other proposals showed that various geometric statements were equivalent to the Euclidean postulate V. It is extremely important that these scholars established the mutual connection between this postulate and the sum of the angles of a triangle and a quadrangle. By their works on the theory of parallel lines Arab mathematicians directly influenced the relevant investigations of their European counterparts. The first European attempt to prove the postulate on parallel lines—made by Witelo, the Polish scientists of the 13th century, while revising Ibn al-Haytham's *Book of Optics*[203] (*Kitab al-Manazir*)—was undoubtedly prompted by Arabic sources. The proofs put forward in the 14th century by the Jewish scholar Levi ben Gerson, who lived in southern France, and by the above-mentioned Alfonso from Spain directly border on Ibn al-Haytham's demonstration. Above, we have demonstrated that *Pseudo-Tusi's Exposition of Euclid* had stimulated both J. Wallis's and G. Saccheri's studies of the theory of parallel lines."

---

203. https://en.m.wikipedia.org/wiki/Book_of_Optics

51.^ ***a***[204] ***b***[205] *Carl B. Boyer*[206] *(2012). History of Analytic Geometry*[207]*. Courier Corporation. ISBN*[208] *978-0-486-15451-0*[209]*. Archived*[210] *from the original on 26 December 2019. Retrieved 18 September2019.*

52.^[211] *C. H. Edwards Jr. (2012). The Historical Development of the Calculus*[212]*. Springer Science & Business Media. p. 95. ISBN*[213] *978-1-4612-6230-5*[214]*. Archived*[215] *from the original on 29 December 2019. Retrieved 18 September2019.*

53.^[216] *Judith V. Field*[217]*; Jeremy Gray (2012). The Geometrical Work of Girard Desargues*[218]*. Springer Science & Business Media. p. 43. ISBN*[219] *978-1-4613-8692-6*[220]*. Archived*[221] *from the original on 27 December 2019. Retrieved 18 September 2019.*

---

204. https://en.m.wikipedia.org/wiki/Geometry#cite_ref-Boyer2012_32-0

205. https://en.m.wikipedia.org/wiki/Geometry#cite_ref-Boyer2012_32-1

206. https://en.m.wikipedia.org/wiki/Carl_Benjamin_Boyer

207. https://books.google.com/books?id=2T4i5fXZbOYC

208. https://en.m.wikipedia.org/wiki/ISBN_(identifier)

209. https://en.m.wikipedia.org/wiki/Special:BookSources/978-0-486-15451-0

210. https://web.archive.org/web/20191226215605/https:/books.google.com/books?id=2T4i5fXZbOYC

211. https://en.m.wikipedia.org/wiki/Geometry#cite_ref-Edwards2012_33-0

212. https://books.google.com/books?id=ilrlBwAAQBAJ&pg=PA95

213. https://en.m.wikipedia.org/wiki/ISBN_(identifier)

214. https://en.m.wikipedia.org/wiki/Special:BookSources/978-1-4612-6230-5

215. https://web.archive.org/web/20191229201529/https:/books.google.com/books?id=ilrlBwAAQBAJ&pg=PA95

216. https://en.m.wikipedia.org/wiki/Geometry#cite_ref-FieldGray2012_34-0

217. https://en.m.wikipedia.org/wiki/Judith_V._Field

218. https://books.google.com/books?id=zSvSBwAAQBAJ&pg=PA43

219. https://en.m.wikipedia.org/wiki/ISBN_(identifier)

220. https://en.m.wikipedia.org/wiki/Special:BookSources/978-1-4613-8692-6

221. https://web.archive.org/web/20191227054645/https:/books.google.com/books?id=zSvSBwAAQBAJ&pg=PA43

54.^[222] *C. R. Wylie (2011). Introduction to Projective Geometry*[223]*. Courier Corporation. ISBN*[224] *978-0-486-14170-1*[225]*. Archived*[226] *from the original on 28 December 2019. Retrieved 18 September2019.*

55.^[227] *Jeremy Gray (2011). Worlds Out of Nothing: A Course in the History of Geometry in the 19th Century*[228]*. Springer Science & Business Media. ISBN*[229] *978-0-85729-060-1*[230]*. Archived*[231] *from the original on 7 December 2019. Retrieved 18 September 2019.*

56.^[232] *Eduardo Bayro-Corrochano (2018). Geometric Algebra Applications Vol. I: Computer Vision, Graphics and Neurocomputing*[233]*. Springer. p. 4. ISBN*[234] *978-3-319-74830-6*[235]*. Archived*[236] *from the original on 28 December 2019. Retrieved 18 September2019.*

57.^[237] *Lôr, Elen Andrea Janzen. "Postulados do Desenho Geométrico"*[238] *(PDF). Universidade Federal do Paraná (in Portuguese). Archived from the original*[239] *(PDF) on 6 February 2015. Retrieved 6 July 2013.*

---

222. https://en.m.wikipedia.org/wiki/Geometry#cite_ref-Wylie2011_35-0

223. https://books.google.com/books?id=VVvGc8kaajEC

224. https://en.m.wikipedia.org/wiki/ISBN_(identifier)

225. https://en.m.wikipedia.org/wiki/Special:BookSources/978-0-486-14170-1

226. https://web.archive.org/web/20191228051716/https:/books.google.com/books?id=VVvGc8kaajEC

227. https://en.m.wikipedia.org/wiki/Geometry#cite_ref-Gray2011_36-0

228. https://books.google.com/books?id=3UeSCvazV0QC

229. https://en.m.wikipedia.org/wiki/ISBN_(identifier)

230. https://en.m.wikipedia.org/wiki/Special:BookSources/978-0-85729-060-1

231. https://web.archive.org/web/20191207041658/https:/books.google.com/books?id=3UeSCvazV0QC

232. https://en.m.wikipedia.org/wiki/Geometry#cite_ref-Bayro-Corrochano2018_37-0

233. https://books.google.com/books?id=SSVhDwAAQBAJ&pg=PA4

234. https://en.m.wikipedia.org/wiki/ISBN_(identifier)

235. https://en.m.wikipedia.org/wiki/Special:BookSources/978-3-319-74830-6

236. https://web.archive.org/web/20191228052142/https:/books.google.com/books?id=SSVhDwAAQBAJ&pg=PA4

237. https://en.m.wikipedia.org/wiki/Geometric_drawing#cite_ref-1

238. https://web.archive.org/web/20150206135137/http:/www.degraf.ufpr.br/docentes/Elen/Desenho%20Geometrico%20Arquitetura.pdf

239. http://www.degraf.ufpr.br/docentes/Elen/Desenho%20Geometrico%20Arquitetura.pdf

58.^[240] *Putnoki, José Carlos. "Que se devolvam a Euclides a régua e compasso". Revista do Professor de Matemática.* **2** *(13). Sociedade Brasileira de Matemática São Paulo: Associação Palas Athena do Brasil: 13–17.*

59.^[241] *"Elenice de Souza Lodron Zuin, Da Régua e do Compasso: As Construções Geométricas Como Um Saber Escolar No Brasil"*[242] *(PDF) (in Portuguese). Archived from the original*[243] *(PDF) on 2014-05-19. Retrieved 2023-02-22.*

60.^[244] *Giongo, Afonso Rocha (1954). Curso de Desenho Geométrico (in Portuguese). São Paulo: Nobel. p. 7.*

61.^[245] *Mandarino, Denis (2007). Desenho Geométrico, construções com régua e compasso (in Portuguese). São Paulo: Plêiade. p. 7.*

62.^ ***a***[246] ***b***[247] *Kern, Willis F.; Bland, James R. (1934). Solid Mensuration With Proofs. New York: John Wiley & Sons.*

63.^[248] *Menn, Stephen (2015). "How Archytas doubled the cube". In Holmes, Brooke; Fischer, Klaus-Dietrich (eds.). The Frontiers of Ancient Science: Essays in honor of Heinrich von Staden. Beiträge zur Altertumskunde. Vol. 338. De Gruyter. pp. 407–436. doi*[249]*:10.1515/9783110336337-021*[250].

64.^[251] *Masià, Ramon (2016). "A new reading of Archytas' doubling of the cube and its implications". Archive for History of Exact Sciences.* **70** *(2): 175–204. doi*[252]*:10.1007/s00407-015-0165-9*[253]*. MR*[254] *3458183*[255].

---

240. https://en.m.wikipedia.org/wiki/Geometric_drawing#cite_ref-2

241. https://en.m.wikipedia.org/wiki/Geometric_drawing#cite_ref-3

242. *https://web.archive.org/web/20140519203412/http:/www.bibliotecadigital.ufmg.br/dspace/bitstream/handle/1843/FAEC-85DGQB/zuin_elenice_disserta_nopw.pdf?sequence=1*

243. *http://www.bibliotecadigital.ufmg.br/dspace/bitstream/handle/1843/FAEC-85DGQB/zuin_elenice_disserta_nopw.pdf?sequence=1*

244. https://en.m.wikipedia.org/wiki/Geometric_drawing#cite_ref-4

245. https://en.m.wikipedia.org/wiki/Geometric_drawing#cite_ref-5

246. ***https://en.m.wikipedia.org/wiki/Doubling_the_cube#cite_ref-Kern1934_2-0***

247. ***https://en.m.wikipedia.org/wiki/Doubling_the_cube#cite_ref-Kern1934_2-1***

248. https://en.m.wikipedia.org/wiki/Doubling_the_cube#cite_ref-3

249. *https://en.m.wikipedia.org/wiki/Doi_(identifier)*

250. *https://doi.org/10.1515%2F9783110336337-021*

251. https://en.m.wikipedia.org/wiki/Doubling_the_cube#cite_ref-4

252. *https://en.m.wikipedia.org/wiki/Doi_(identifier)*

65.^[256] *Guilbeau, Lucye (1930). "The history of the solution of the cubic equation". Mathematics News Letter.* **5** *(4): 8–12. doi*[257]*:10.2307/3027812*[258]*. JSTOR*[259] *3027812*[260]*.*

66.^[261] *Zhmud', Leonid I͡Akovlevich (2006). The Origin of the History of Science in Classical Antiquity*[262]*. Walter de Gruyter. pp. 84, quoting Plutarch and Theon of Smyrna. ISBN*[263] *978-3-11-017966-8*[264]*.*

67.^ *"Plutarch, De E apud Delphos, section 6 386.4"*[265]*. www.perseus.tufts.edu. Retrieved 2024-09-17.*

68.^[266] Plutarch[267], De genio Socratis 579.B

69.^[268] (Plut., *Quaestiones convivales* VIII.ii[269], 718ef)

70.^[270] Carl Werner Müller, *Die Kurzdialoge der Appendix Platonica,* Munich: Wilhelm Fink, 1975, pp. 105–106

---

253. *https://doi.org/10.1007%2Fs00407-015-0165-9*

254. *https://en.m.wikipedia.org/wiki/MR_(identifier)*

255. *https://mathscinet.ams.org/mathscinet-getitem?mr=3458183*

256. https://en.m.wikipedia.org/wiki/Doubling_the_cube#cite_ref-5

257. *https://en.m.wikipedia.org/wiki/Doi_(identifier)*

258. *https://doi.org/10.2307%2F3027812*

259. *https://en.m.wikipedia.org/wiki/JSTOR_(identifier)*

260. *https://www.jstor.org/stable/3027812*

261. https://en.m.wikipedia.org/wiki/Doubling_the_cube#cite_ref-Zhmud_6-0

262. *https://books.google.com/books?id=oX28qf7LKdoC&pg=PA84*

263. *https://en.m.wikipedia.org/wiki/ISBN_(identifier)*

264. *https://en.m.wikipedia.org/wiki/Special:BookSources/978-3-11-017966-8*

265. *https://www.perseus.tufts.edu/hopper/text?doc=Perseus:text:2008.01.0243:section=6*

266. https://en.m.wikipedia.org/wiki/Doubling_the_cube#cite_ref-8

267. https://en.m.wikipedia.org/wiki/Plutarch

268. https://en.m.wikipedia.org/wiki/Doubling_the_cube#cite_ref-9

269. https://web.archive.org/web/20080815001514/http:/ebooks.adelaide.edu.au/p/plutarch/symposiacs/chapter8.html#section80

270. https://en.m.wikipedia.org/wiki/Doubling_the_cube#cite_ref-10

71.^[271] *Knorr, Wilbur Richard*[272] *(1986), The Ancient Tradition of Geometric Problems*[273]*, Dover Books on Mathematics, Courier Dover Publications, p. 4*[274]*, ISBN*[275] *9780486675329*[276]*.*

72.^[277] T.L. Heath *A History of Greek Mathematics*[278], Vol. 1

73.^[279] *Lützen, Jesper (24 January 2010). "The Algebra of Geometric Impossibility: Descartes and Montucla on the Impossibility of the Duplication of the Cube and the Trisection of the Angle"*[280]*. Centaurus.* ***52*** *(1): 4–37. doi*[281]*:10.1111/j.1600-0498.2009.00160.x*[282]*.*

74.^[283] *Knorr, Wilbur Richard*[284] *(1989). "Pappus' texts on cube duplication". Textual Studies in Ancient and Medieval Geometry*[285]*. Boston: Birkhäuser. pp. 63–76*[286]*. doi*[287]*:10.1007/978-1-4612-3690-0_5*[288]*. ISBN*[289] *9780817633875*[290]*.*

75.^[291] *Dörrie, Heinrich (1965). 100 Great Problems of Elementary Mathematics. Dover. p. 171. ISBN*[292] *0486-61348-8*[293]*.*

---

271. **https://en.m.wikipedia.org/wiki/Doubling_the_cube#cite_ref-11**

272. *https://en.m.wikipedia.org/wiki/Wilbur_Knorr*

273. *https://en.m.wikipedia.org/wiki/The_Ancient_Tradition_of_Geometric_Problems*

274. *https://books.google.com/books?id=sgQS5BaTWl4C&pg=PA4*

275. *https://en.m.wikipedia.org/wiki/ISBN_(identifier)*

276. *https://en.m.wikipedia.org/wiki/Special:BookSources/9780486675329*

277. **https://en.m.wikipedia.org/wiki/Doubling_the_cube#cite_ref-Heath_12-0**

278. *https://en.m.wikipedia.org/wiki/A_History_of_Greek_Mathematics*

279. **https://en.m.wikipedia.org/wiki/Doubling_the_cube#cite_ref-13**

280. *https://onlinelibrary.wiley.com/doi/10.1111/j.1600-0498.2009.00160.x*

281. *https://en.m.wikipedia.org/wiki/Doi_(identifier)*

282. *https://doi.org/10.1111%2Fj.1600-0498.2009.00160.x*

283. **https://en.m.wikipedia.org/wiki/Doubling_the_cube#cite_ref-14**

284. *https://en.m.wikipedia.org/wiki/Wilbur_Knorr*

285. *https://archive.org/details/textualstudiesin0000knor*

286. *https://archive.org/details/textualstudiesin0000knor/page/63*

287. *https://en.m.wikipedia.org/wiki/Doi_(identifier)*

288. *https://doi.org/10.1007%2F978-1-4612-3690-0_5*

289. *https://en.m.wikipedia.org/wiki/ISBN_(identifier)*

290. *https://en.m.wikipedia.org/wiki/Special:BookSources/9780817633875*

291. **https://en.m.wikipedia.org/wiki/Doubling_the_cube#cite_ref-15**

292. *https://en.m.wikipedia.org/wiki/ISBN_(identifier)*

76.Carl B Boyer, *A history of mathematics*, john Wiley and sons, 1968, pg 69-176.

77.E T Bell, *The development of mathematics,* Mc-Graw Hill book company, New York, London, 1945, pg 76-79.

78.H Eves, *An introduction to the history of mathematics, 5th edition,* Sounders, New York, 1983.

79.J H Smith, *Elements of geometry,* Irvington, London, Oxford, Cambridge, 1871.

80.R Courant and H Robbins, *What is mathematics?* Oxford university press, paperback edition, 1978, pg 120-135.

81.Sir T L Heath, *A manual of Greek mathematics,* Oxford university press, 1963, pg 154-170.

---

293. *https://en.m.wikipedia.org/wiki/Special:BookSources/0486-61348-8*

[1] The **history of painting** reaches back in time to artifacts and artwork created by pre-historic artists, and spans all cultures. It represents a continuous, though periodically disrupted, tradition from Antiquity. Across cultures, continents, and millennia, the history of painting consists of an ongoing river of creativity that continues into the 21st century.[1][294] Until the early 20th century it relied primarily on representational[295], religious[296] and classical[297] motifs, after which time more purely abstract[298] and conceptual[299] approaches gained favor.

History of painting

[300]

294. https://en.m.wikipedia.org/wiki/History_of_painting#cite_note-ColeGealt1991-1
295. https://en.m.wikipedia.org/wiki/Representational_art
296. https://en.m.wikipedia.org/wiki/Religious_art
297. https://en.m.wikipedia.org/wiki/Classical_antiquity
298. https://en.m.wikipedia.org/wiki/Abstract_art
299. https://en.m.wikipedia.org/wiki/Conceptual_art
300. https://en.m.wikipedia.org/wiki/File:Lubang_Jeriji_Sal%C3%A9h_cave_painting_of_Bull.jpg

Lubang Jeriji Saléh[301] cave, in Kalimantan[302], Indonesia[303], contains one of the oldest known figurative paintings[304], a 40,000-year-old depiction of a bull.

305

The Milkmaid[306] by Johannes Vermeer[307], c. 1657

Developments in Eastern painting historically parallel those in Western painting[308], in general, a few centuries earlier.[2][309]

---

301. https://en.m.wikipedia.org/wiki/Lubang_Jeriji_Sal%C3%A9h

302. https://en.m.wikipedia.org/wiki/Kalimantan

303. https://en.m.wikipedia.org/wiki/Indonesia

304. https://en.m.wikipedia.org/wiki/Figurative_art

305. https://en.m.wikipedia.org/wiki/File:Johannes_Vermeer_-_Het_melkmeisje_-_Google_Art_Project.jpg

306. https://en.m.wikipedia.org/wiki/The_Milkmaid_(Vermeer)

307. https://en.m.wikipedia.org/wiki/Johannes_Vermeer

African art[310], Jewish art[311], Islamic art[312], Indonesian art[313], Indian art[314],[3][315]Chinese art[316], and Japanese art[317][4][318] each had significant influence on Western art, and vice versa.[5][319]

Initially serving utilitarian purpose, followed by imperial, private, civic, and religious patronage, Eastern and Western painting later found audiences in the aristocracy[320] and the middle class. From the Modern era, the Middle Ages[321] through the Renaissance painters worked for the church and a wealthy aristocracy.[6][322] Beginning with the Baroque[323] era artists received private commissions from a more educated and prosperous middle class.[7][324] Finally in the West the idea of

308. https://en.m.wikipedia.org/wiki/Western_painting

309. https://en.m.wikipedia.org/wiki/History_of_painting#cite_note-ReferenceA-2

310. https://en.m.wikipedia.org/wiki/African_art

311. https://en.m.wikipedia.org/wiki/Jewish_art

312. https://en.m.wikipedia.org/wiki/Islamic_art

313. https://en.m.wikipedia.org/wiki/Indonesian_art

314. https://en.m.wikipedia.org/wiki/Indian_art

315. https://en.m.wikipedia.org/wiki/History_of_painting#cite_note-3

316. https://en.m.wikipedia.org/wiki/Chinese_art

317. https://en.m.wikipedia.org/wiki/Japanese_art

318. https://en.m.wikipedia.org/wiki/History_of_painting#cite_note-Publisher_1999-4

319. https://en.m.wikipedia.org/wiki/History_of_painting#cite_note-Western_Art_1989-5

320. https://en.m.wikipedia.org/wiki/Aristocracy

321. https://en.m.wikipedia.org/wiki/Middle_Ages

322. https://en.m.wikipedia.org/wiki/History_of_painting#cite_note-6

323. https://en.m.wikipedia.org/wiki/Baroque

"art for art's sake[325]"[8][326] began to find expression in the work of the Romantic[327] painters like Francisco de Goya[328], John Constable[329], and J. M. W. Turner[330].[9][331] The 19th century saw the rise of the commercial art gallery[332], which provided patronage in the 20th century.[10][333]

324. https://en.m.wikipedia.org/wiki/History_of_painting#cite_note-7
325. https://en.m.wikipedia.org/wiki/Art_for_art%27s_sake
326. https://en.m.wikipedia.org/wiki/History_of_painting#cite_note-8
327. https://en.m.wikipedia.org/wiki/Romanticism
328. https://en.m.wikipedia.org/wiki/Francisco_de_Goya
329. https://en.m.wikipedia.org/wiki/John_Constable
330. https://en.m.wikipedia.org/wiki/J._M._W._Turner
331. https://en.m.wikipedia.org/wiki/History_of_painting#cite_note-9
332. https://en.m.wikipedia.org/wiki/Art_gallery
333. https://en.m.wikipedia.org/wiki/History_of_painting#cite_note-10

[2] Beginning of the Stone Age

O

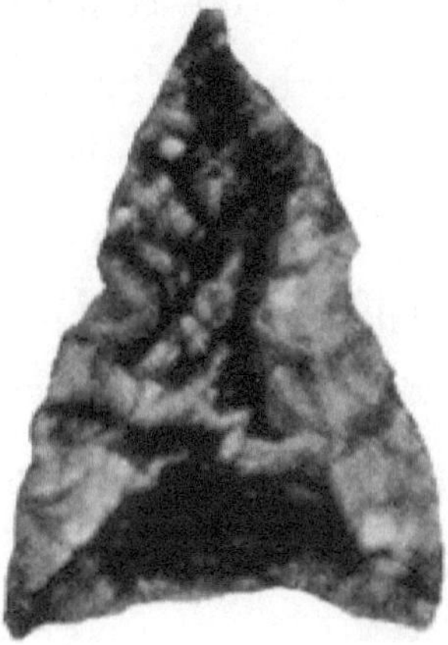

[1]bsidian[2] projectile point[3]

The oldest indirect evidence found of stone tool use is fossilized animal bones with tool marks; these are 3.4 million years old and were found in the Lower Awash Valley in Ethiopia.[13][4] Archaeological discoveries in Kenya in 2015, identifying what may be the oldest evidence of hominin use of

1. https://en.m.wikipedia.org/wiki/File:Arrowhead.jpg

2. https://en.m.wikipedia.org/wiki/Obsidian

3. https://en.m.wikipedia.org/wiki/Projectile_point

4. https://en.m.wikipedia.org/wiki/Stone_Age#cite_note-nhm.ac.uk-1

tools known to date, have indicated that *Kenyanthropus*[5] *platyops* (a 3.2 to 3.5-million-year-old Pliocene[6] hominin fossil discovered in Lake Turkana, Kenya, in 1999) may have been the earliest tool-users known.[19][7]

The oldest stone tools were excavated from the site of Lomekwi[8] 3 in West Turkana[9], northwestern Kenya, and date to 3.3 million years old.[20][10] Prior to the discovery of these "Lomekwian" tools, the oldest known stone tools had been found at several sites at Gona, Ethiopia[11], on sediments of the paleo-Awash River[12], which serve to date them. All the tools come from the Busidama Formation, which lies above a disconformity[13], or missing layer, which would have been from 2.9 to 2.7 mya[14]. The oldest sites discovered to contain tools are dated to 2.6–2.55 mya.[21][15] One of the most striking circumstances about these sites is that they are from the Late Pliocene[16], where prior to their

5. *https://en.m.wikipedia.org/wiki/Kenyanthropus*
6. https://en.m.wikipedia.org/wiki/Pliocene
7. https://en.m.wikipedia.org/wiki/Stone_Age#cite_note-7
8. https://en.m.wikipedia.org/wiki/Lomekwi
9. https://en.m.wikipedia.org/wiki/Turkana_County
10. https://en.m.wikipedia.org/wiki/Stone_Age#cite_note-Harmand_2015-8
11. https://en.m.wikipedia.org/wiki/Gona,_Ethiopia
12. https://en.m.wikipedia.org/wiki/Awash_River
13. https://en.m.wikipedia.org/wiki/Disconformity
14. https://en.m.wikipedia.org/wiki/Mya_(unit)
15. https://en.m.wikipedia.org/wiki/Stone_Age#cite_note-9
16. https://en.m.wikipedia.org/wiki/Pliocene

discovery tools were thought to have evolved only in the Pleistocene[17]. Excavators at the locality point out that:[22][18]

> ... the earliest stone tool makers were skilled flintknappers[19] ... The possible reasons behind this seeming abrupt transition from the absence of stone tools to the presence thereof include ... gaps in the geological record.

The species that made the Pliocene tools remains unknown. Fragments of *Australopithecus garhi*[20], *Australopithecus aethiopicus*[21],[23][22] and *Homo*, possibly *Homo habilis*[23], have been found in sites near the age of the Gona tools.[24][24]

In July 2018, scientists reported the discovery in China[25] of the known oldest stone tools outside Africa, estimated at 2.12 million years old.[25][26]

[3] *Main article: History of geometry*[27]

---

17. https://en.m.wikipedia.org/wiki/Pleistocene

18. https://en.m.wikipedia.org/wiki/Stone_Age#cite_note-10

19. https://en.m.wikipedia.org/wiki/Flintknappers

20. *https://en.m.wikipedia.org/wiki/Australopithecus_garhi*

21. *https://en.m.wikipedia.org/wiki/Australopithecus_aethiopicus*

22. https://en.m.wikipedia.org/wiki/Stone_Age#cite_note-11

23. *https://en.m.wikipedia.org/wiki/Homo_habilis*

24. https://en.m.wikipedia.org/wiki/Stone_Age#cite_note-12

25. https://en.m.wikipedia.org/wiki/China

26. https://en.m.wikipedia.org/wiki/Stone_Age#cite_note-NYT-20180711cz-13

27. *https://en.m.wikipedia.org/wiki/History_of_geometry*

[28]A European[29] and an Arab[30] practicing geometry in the 15th century

The earliest recorded beginnings of geometry can be traced to ancient Mesopotamia[31] and Egypt[32] in the 2nd millennium BC.26][33][27][34] Early geometry was a collection of empirically discovered principles concerning lengths, angles, areas, and volumes, which were developed to meet some practical need in surveying[35], construction[36], astronomy[37], and various crafts. The earliest known texts on geometry are the Egyptian[38] Rhind

28. https://en.m.wikipedia.org/wiki/ File:Westerner_and_Arab_practicing_geometry_15th_century_manuscript.jpg
29. https://en.m.wikipedia.org/wiki/Ethnic_groups_in_Europe
30. https://en.m.wikipedia.org/wiki/Arab
31. https://en.m.wikipedia.org/wiki/Mesopotamia
32. https://en.m.wikipedia.org/wiki/Ancient_Egypt
33. https://en.m.wikipedia.org/wiki/Geometry#cite_note-6
34. https://en.m.wikipedia.org/wiki/Geometry#cite_note-7
35. https://en.m.wikipedia.org/wiki/Surveying
36. https://en.m.wikipedia.org/wiki/Construction
37. https://en.m.wikipedia.org/wiki/Astronomy

Papyrus[39](2000–1800 BC) and Moscow Papyrus[40](c. 1890 BC), and the Babylonian clay tablets[41], such as Plimpton 322[42] (1900 BC). For example, the Moscow Papyrus gives a formula for calculating the volume of a truncated pyramid, or frustum[43].[28][44] Later clay tablets (350–50 BC) demonstrate that Babylonian astronomers implemented trapezoid[45] procedures for computing Jupiter's position and motion[46] within time-velocity space. These geometric procedures anticipated the Oxford Calculators[47], including the mean speed theorem[48], by 14 centuries.[29][49]South of Egypt the Nubians established a system of geometry including early versions of sun clocks.[30][50][31][51]

---

38. https://en.m.wikipedia.org/wiki/Egyptian_mathematics
39. https://en.m.wikipedia.org/wiki/Rhind_Mathematical_Papyrus
40. https://en.m.wikipedia.org/wiki/Moscow_Mathematical_Papyrus
41. https://en.m.wikipedia.org/wiki/Babylonian_mathematics
42. https://en.m.wikipedia.org/wiki/Plimpton_322
43. https://en.m.wikipedia.org/wiki/Frustum
44. https://en.m.wikipedia.org/wiki/Geometry#cite_note-Boyer_1991_loc_43ec3e5dee6e706af7766fffea512721_Egypt_p._19-8
45. https://en.m.wikipedia.org/wiki/Trapezoid
46. https://en.m.wikipedia.org/wiki/Displacement_(vector)
47. https://en.m.wikipedia.org/wiki/Oxford_Calculators
48. https://en.m.wikipedia.org/wiki/Mean_speed_theorem
49. https://en.m.wikipedia.org/wiki/Geometry#cite_note-9
50. https://en.m.wikipedia.org/wiki/Geometry#cite_note-10
51. https://en.m.wikipedia.org/wiki/Geometry#cite_note-11

In the 7th century BC, the Greek mathematician Thales of Miletus[52] used geometry to solve problems such as calculating the height of pyramids and the distance of ships from the shore. He is credited with the first use of deductive reasoning applied to geometry, by deriving four corollaries to Thales's theorem[53].[32][54]Pythagoras[55] established the Pythagorean School[56], which is credited with the first proof of the Pythagorean theorem[57],[33][58] though the statement of the theorem has a long history.[34][59][35][60] Eudoxus[61] (408–c. 355 BC) developed the method of exhaustion[62], which allowed the calculation of areas and volumes of curvilinear figures,[36][63] as well as a theory of ratios that avoided the problem of incommensurable magnitudes[64], which enabled subsequent geometers to make significant advances. Around 300 BC,

---

52. https://en.m.wikipedia.org/wiki/Thales_of_Miletus
53. https://en.m.wikipedia.org/wiki/Thales%27s_theorem
54. https://en.m.wikipedia.org/wiki/Geometry#cite_note-Boyer_1991_loc_43ec3e5dee6e706af7766fffea512721_Ionia_and_the_Pythagoreans_p._43-12
55. https://en.m.wikipedia.org/wiki/Pythagoras
56. https://en.m.wikipedia.org/wiki/Pythagoreans
57. https://en.m.wikipedia.org/wiki/Pythagorean_theorem
58. https://en.m.wikipedia.org/wiki/Geometry#cite_note-13
59. https://en.m.wikipedia.org/wiki/Geometry#cite_note-14
60. https://en.m.wikipedia.org/wiki/Geometry#cite_note-15
61. https://en.m.wikipedia.org/wiki/Eudoxus_of_Cnidus
62. https://en.m.wikipedia.org/wiki/Method_of_exhaustion
63. https://en.m.wikipedia.org/wiki/Geometry#cite_note-16
64. https://en.m.wikipedia.org/wiki/Incommensurable_magnitudes

geometry was revolutionized by Euclid, whose *Elements*[65], widely considered the most successful and influential textbook of all time,[37][66] introduced mathematical rigor[67] through the axiomatic method[68] and is the earliest example of the format still used in mathematics today, that of definition, axiom, theorem, and proof. Although most of the contents of the *Elements* were already known, Euclid arranged them into a single, coherent logical framework.[38][69] The *Elements* was known to all educated people in the West until the middle of the 20th century and its contents are still taught in geometry classes today.[39][70]Archimedes[71] (c. 287–212 BC) of Syracuse, Italy[72] used the method of exhaustion to calculate the area[73] under the arc of a parabola with the summation of an infinite series[74], and gave remarkably accurate approximations of pi[75].[40][76] He also

---

65. *https://en.m.wikipedia.org/wiki/Euclid%27s_Elements*
66. https://en.m.wikipedia.org/wiki/Geometry#cite_note-17
67. https://en.m.wikipedia.org/wiki/Mathematical_rigor
68. https://en.m.wikipedia.org/wiki/Axiomatic_method
69. https://en.m.wikipedia.org/wiki/Geometry#cite_note-Boyer_1991_loc_43ec3e5dee6e706af7766fffea512721_Euclid_of_Alexandria_p._104-18
70. https://en.m.wikipedia.org/wiki/Geometry#cite_note-19
71. https://en.m.wikipedia.org/wiki/Archimedes
72. https://en.m.wikipedia.org/wiki/Syracuse,_Italy
73. https://en.m.wikipedia.org/wiki/Area
74. https://en.m.wikipedia.org/wiki/Series_(mathematics)
75. https://en.m.wikipedia.org/wiki/Pi
76. https://en.m.wikipedia.org/wiki/Geometry#cite_note-20

studied the spiral[77] bearing his name and obtained formulas for the volumes of surfaces of revolution[78].

[79] *Woman teaching geometry.* Illustration at the beginning of a medieval translation of Euclid's Elements[80], (c. 1310).

Indian[81] mathematicians also made many important contributions in geometry. The *Shatapatha Brahmana*[82] (3rd century BC) contains rules for ritual geometric constructions that are similar to the *Sulba Sutras*[83].[41][84] According to (Hayashi 2005[85], p. 363), the *Śulba Sūtras* contain "the earliest extant verbal expression of the Pythagorean Theorem in the

---

77. https://en.m.wikipedia.org/wiki/Archimedes_spiral
78. https://en.m.wikipedia.org/wiki/Surface_of_revolution
79. https://en.m.wikipedia.org/wiki/File:Woman_teaching_geometry.jpg
80. https://en.m.wikipedia.org/wiki/Euclid%27s_Elements
81. https://en.m.wikipedia.org/wiki/Indian_mathematics
82. *https://en.m.wikipedia.org/wiki/Shatapatha_Brahmana*
83. *https://en.m.wikipedia.org/wiki/Shulba_Sutras*
84. https://en.m.wikipedia.org/wiki/Geometry#cite_note-Staal_1999-21
85. https://en.m.wikipedia.org/wiki/Geometry#CITEREFHayashi2005

world, although it had already been known to the Old Babylonians. They contain lists of Pythagorean triples[86],[b][87] which are particular cases of Diophantine equations[88].[42][89] In the Bakhshali manuscript[90], there are a handful of geometric problems (including problems about volumes of irregular solids). The Bakhshali manuscript also "employs a decimal place value system with a dot for zero."[43][91] Aryabhata[92]'s *Aryabhatiya*[93] (499) includes the computation of areas and volumes. Brahmagupta[94] wrote his astronomical work *Brāhmasphuṭasiddhānta*[95] in 628. Chapter 12, containing 66 Sanskrit verses, was divided into two sections: "basic operations" (including cube roots, fractions, ratio and proportion, and barter) and "practical mathematics" (including mixture, mathematical series, plane figures, stacking bricks, sawing of timber, and piling of grain).[44][96] In the latter section, he stated his famous theorem on the diagonals of a cyclic quadrilateral[97].

---

86. https://en.m.wikipedia.org/wiki/Pythagorean_triples
87. https://en.m.wikipedia.org/wiki/Geometry#cite_note-22
88. https://en.m.wikipedia.org/wiki/Diophantine_equations
89. https://en.m.wikipedia.org/wiki/Geometry#cite_note-cooke198-23
90. https://en.m.wikipedia.org/wiki/Bakhshali_manuscript
91. https://en.m.wikipedia.org/wiki/Geometry#cite_note-hayashi2005-371-24
92. https://en.m.wikipedia.org/wiki/Aryabhata
93. *https://en.m.wikipedia.org/wiki/Aryabhatiya*
94. https://en.m.wikipedia.org/wiki/Brahmagupta
95. *https://en.m.wikipedia.org/wiki/Brahmasphutasiddhanta*
96. https://en.m.wikipedia.org/wiki/Geometry#cite_note-hayashi2003-p121-122-25
97. https://en.m.wikipedia.org/wiki/Cyclic_quadrilateral

Chapter 12 also included a formula for the area of a cyclic quadrilateral (a generalization of Heron's formula[98]), as well as a complete description of rational triangles[99](*i.e.* triangles with rational sides and rational areas).[45][100]

In the Middle Ages[101], mathematics in medieval Islam[102] contributed to the development of geometry, especially algebraic geometry[103].[46][104][47][105] Al-Mahani[106] (b. 853) conceived the idea of reducing geometrical problems such as duplicating the cube to problems in algebra.[48][107] Thābit ibn Qurra[108] (known as Thebit in Latin[109]) (836–901) dealt with arithmetic operations applied to ratios[110] of geometrical quantities, and contributed to the development of analytic geometry[111].[49][112] Omar Khayyam[113] (1048–1131) found geometric solutions to

98. https://en.m.wikipedia.org/wiki/Heron%27s_formula
99. https://en.m.wikipedia.org/wiki/Rational_triangle
100. https://en.m.wikipedia.org/wiki/Geometry#cite_note-hayashi2003-p121-122-25
101. https://en.m.wikipedia.org/wiki/Middle_Ages
102. https://en.m.wikipedia.org/wiki/Mathematics_in_medieval_Islam
103. https://en.m.wikipedia.org/wiki/Algebraic_geometry
104. https://en.m.wikipedia.org/wiki/Geometry#cite_note-26
105. https://en.m.wikipedia.org/wiki/Geometry#cite_note-27
106. https://en.m.wikipedia.org/wiki/Al-Mahani
107. https://en.m.wikipedia.org/wiki/Geometry#cite_note-28
108. https://en.m.wikipedia.org/wiki/Th%C4%81bit_ibn_Qurra
109. https://en.m.wikipedia.org/wiki/Latin
110. https://en.m.wikipedia.org/wiki/Ratio
111. https://en.m.wikipedia.org/wiki/Analytic_geometry
112. https://en.m.wikipedia.org/wiki/Geometry#cite_note-ReferenceA-29

cubic equations[114].[50][115] The theorems of Ibn al-Haytham[116] (Alhazen), Omar Khayyam and Nasir al-Din al-Tusi[117] on quadrilaterals[118], including the Lambert quadrilateral[119] and Saccheri quadrilateral[120], were part of a line of research on the parallel postulate[121] continued by later European geometers, including Vitello[122](c. 1230 – c. 1314), Gersonides[123] (1288–1344), Alfonso, John Wallis[124], and Giovanni Girolamo Saccheri[125], that by the 19th century led to the discovery of hyperbolic geometry[126].[51][127]

In the early 17th century, there were two important developments in geometry. The first was the creation of analytic geometry, or geometry with coordinates[128] and equations[129], by

---

113. https://en.m.wikipedia.org/wiki/Omar_Khayyam
114. https://en.m.wikipedia.org/wiki/Cubic_equation
115. https://en.m.wikipedia.org/wiki/Geometry#cite_note-30
116. https://en.m.wikipedia.org/wiki/Ibn_al-Haytham
117. https://en.m.wikipedia.org/wiki/Nasir_al-Din_al-Tusi
118. https://en.m.wikipedia.org/wiki/Quadrilateral
119. https://en.m.wikipedia.org/wiki/Lambert_quadrilateral
120. https://en.m.wikipedia.org/wiki/Saccheri_quadrilateral
121. https://en.m.wikipedia.org/wiki/Parallel_postulate
122. https://en.m.wikipedia.org/wiki/Vitello
123. https://en.m.wikipedia.org/wiki/Gersonides
124. https://en.m.wikipedia.org/wiki/John_Wallis
125. https://en.m.wikipedia.org/wiki/Giovanni_Girolamo_Saccheri
126. https://en.m.wikipedia.org/wiki/Hyperbolic_geometry
127. https://en.m.wikipedia.org/wiki/Geometry#cite_note-31
128. https://en.m.wikipedia.org/wiki/Coordinate_system
129. https://en.m.wikipedia.org/wiki/Equation

René Descartes[130] (1596–1650) and Pierre de Fermat[131] (1601–1665).[52][132] This was a necessary precursor to the development of calculus[133] and a precise quantitative science of physics[134].[53][135] The second geometric development of this period was the systematic study of projective geometry[136] by Girard Desargues[137] (1591–1661).[54][138] Projective geometry studies properties of shapes which are unchanged under projections[139] and sections[140], especially as they relate to artistic perspective[141].[55][142]

Two developments in geometry in the 19th century changed the way it had been studied previously.[56][143] These were the discovery of non-Euclidean geometries[144] by Nikolai Ivanovich Lobachevsky, János Bolyai and Carl Friedrich Gauss and of the formulation of symmetry[145] as the central consideration in the

130. https://en.m.wikipedia.org/wiki/Ren%C3%A9_Descartes
131. https://en.m.wikipedia.org/wiki/Pierre_de_Fermat
132. https://en.m.wikipedia.org/wiki/Geometry#cite_note-Boyer2012-32
133. https://en.m.wikipedia.org/wiki/Calculus
134. https://en.m.wikipedia.org/wiki/Physics
135. https://en.m.wikipedia.org/wiki/Geometry#cite_note-Edwards2012-33
136. https://en.m.wikipedia.org/wiki/Projective_geometry
137. https://en.m.wikipedia.org/wiki/Girard_Desargues
138. https://en.m.wikipedia.org/wiki/Geometry#cite_note-FieldGray2012-34
139. https://en.m.wikipedia.org/wiki/Projection_(linear_algebra)
140. https://en.m.wikipedia.org/wiki/Section_(fiber_bundle)
141. https://en.m.wikipedia.org/wiki/Perspective_(graphical)
142. https://en.m.wikipedia.org/wiki/Geometry#cite_note-Wylie2011-35
143. https://en.m.wikipedia.org/wiki/Geometry#cite_note-Gray2011-36
144. https://en.m.wikipedia.org/wiki/Non-Euclidean_geometry

Erlangen programme[146] of Felix Klein[147] (which generalized the Euclidean and non-Euclidean geometries). Two of the master geometers of the time were Bernhard Riemann[148] (1826–1866), working primarily with tools from mathematical analysis[149], and introducing the Riemann surface[150], and Henri Poincaré[151], the founder of algebraic topology[152] and the geometric theory of dynamical systems[153]. As a consequence of these major changes in the conception of geometry, the concept of "space[154]" became something rich and varied, and the natural background for theories as different as complex analysis[155] and classical mechanics[156].[35][157]

[4] **Geometric drawing** consists of a set of processes for constructing geometric shapes and solving problems with the use of a ruler without graduation and the compass (drawing tool)[158].[57][159][58][160] Modernly, such studies can be done with

---

145. https://en.m.wikipedia.org/wiki/Symmetry
146. https://en.m.wikipedia.org/wiki/Erlangen_programme
147. https://en.m.wikipedia.org/wiki/Felix_Klein
148. https://en.m.wikipedia.org/wiki/Bernhard_Riemann
149. https://en.m.wikipedia.org/wiki/Mathematical_analysis
150. https://en.m.wikipedia.org/wiki/Riemann_surface
151. https://en.m.wikipedia.org/wiki/Henri_Poincar%C3%A9
152. https://en.m.wikipedia.org/wiki/Algebraic_topology
153. https://en.m.wikipedia.org/wiki/Dynamical_system
154. https://en.m.wikipedia.org/wiki/Space_(mathematics)
155. https://en.m.wikipedia.org/wiki/Complex_analysis
156. https://en.m.wikipedia.org/wiki/Classical_mechanics
157. https://en.m.wikipedia.org/wiki/Geometry#cite_note-Bayro-Corrochano2018-37

the aid of software[161], which simulates the strokes performed by these instruments.[59][162]

[163]Geometric drawing made with ruler and compass.

For ancient mathematicians, geometry[164] could not do without the methods of geometric constructions, necessary for understanding, theoretical enrichment, and problem-solving.

The accuracy and precision required of geometric drawing make it an important ally in the application of geometric concepts in significant areas of human knowledge, such as

158. https://en.m.wikipedia.org/wiki/Compass_(drawing_tool)
159. https://en.m.wikipedia.org/wiki/Geometric_drawing#cite_note-1
160. https://en.m.wikipedia.org/wiki/Geometric_drawing#cite_note-2
161. https://en.m.wikipedia.org/wiki/Cabri_Geometry
162. https://en.m.wikipedia.org/wiki/Geometric_drawing#cite_note-3
163. https://en.m.wikipedia.org/wiki/File:Pesne_Geometrician.jpg
164. https://en.m.wikipedia.org/wiki/Geometry

architecture[165], engineering[166], industrial design[167], among others.

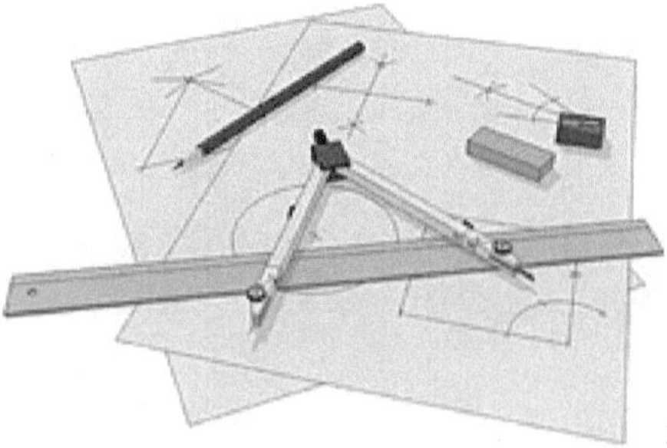

[168]Ruler and compass.

The process of geometric drawing is based on constructions with a ruler and compass, which in turn are based on the first three postulates of Euclid's *Elements*[169].

The historical importance of rulers and compasses as instruments in solving geometric problems leads many authors to limit Geometric Drawing to the representation and solution of geometric figures in the plane.[60][170]

With the development of computer-aided design (CAD[171]) programs, geometric drawing has become more important in teaching-learning processes (development of spatial faculties) than the more imprecise tracing offered by rulers and compasses, when taking into account the precision of computer systems.[61][172]

165. https://en.m.wikipedia.org/wiki/Architecture

166. https://en.m.wikipedia.org/wiki/Engineering

167. https://en.m.wikipedia.org/wiki/Industrial_design

168. https://en.m.wikipedia.org/wiki/File:R%C3%A9gua_e_compasso.jpg

169. https://en.m.wikipedia.org/wiki/Euclid%27s_Elements

170. https://en.m.wikipedia.org/wiki/Geometric_drawing#cite_note-4

171. https://en.m.wikipedia.org/wiki/Computer-aided_design

172. https://en.m.wikipedia.org/wiki/Geometric_drawing#cite_note-5

Don't miss out!

Visit the website below and you can sign up to receive emails whenever manoranjan ghoshal publishes a new book. There's no charge and no obligation.

https://books2read.com/r/B-A-SODHD-HGIXI

Connecting independent readers to independent writers.

www.ingramcontent.com/pod-product-compliance
Lightning Source LLC
LaVergne TN
LVHW090429210926
848480LV00002B/751